Our Universe and How We Got Here

Gerry Krystal, PhD

Copyright © 2023: Gerry Krystal

All rights reserved.

ISBN: 979-8-3982-4108-2

DEDICATION

I would like to dedicate this book to all the scientists who sought the truth throughout our turbulent human history. Some of them, far braver than me, sought the truth even though it meant excommunication or death. I would also like to dedicate this book to the two traits I consider essential for our future survival as a species, kindness and love, in all its myriad forms. If we fully embrace these feelings, I think we can finally warrant the name we have immodestly given ourselves, Homo sapiens (wise people).

CONTENTS

Overview	1
Chapter 1. Our Universe	5
Chapter 2. Earth – A Goldilocks Planet	30
Chapter 3. Life	44
Chapter 4. Evolution	74
Chapter 5. Biogenesis	154
Chapter 6. Final Thoughts	183
References	188
About the Author	194
About the Illustrator/Editor	194

ACKNOWLEDGMENTS

I would very much like to thank my family and friends for their encouragement during the writing of this book. Some encouraged me to keep going while others encouraged me to stop (just kidding) and get back to my research into the effects of diets on cancer. Either way, it's still encouragement and I thank them for it. I would really like to thank my lovely wife for proofreading this manuscript and my daughter, Samantha, for both editing and making the beautiful illustrations, including the doodles and awesome book cover. So here we go! Hope you like it.

OVERVIEW

The more clearly we can focus our attention on the wonders and realities of the universe about us, the less taste we shall have for destruction **Rachel Carson**

Dear Reader,

What I have tried to cover in this book, in a lighthearted, and, I hope, easy-to-understand fashion, is what is currently known about our universe, our "Goldilocks" planet, life in its myriad forms, the critical steps in our evolution, and how life may have started on this planet.

Specifically, in Chapter 1, I touch on how our "current" universe likely began and will likely end. I also explain fusion, which occurs in the center of our stars and is the most important reaction in our universe. As well, I mention how stars not only generate light and heat, but all the elements that make up everything, and so we, and all other living (and non-living) things are truly made of stardust. I then introduce an attractive alternative to the Big Bang Theory (the Big Bounce) and the unimaginable possibility of endless space and time.

In Chapter 2, I demonstrate that a planet is conducive to life (as we know it) if it has an atmosphere, because an atmosphere prevents water, which is crucial for life, from boiling away into space. I also explain how lucky we are that Earth has a molten core, even though a molten core leads to tectonic plates, which, because of their movement, cause volcanoes and earthquakes!

In Chapter 3, I go into how all life is made up of cells and the similarities and differences amongst the different cell types that currently exist. Then, I go on to explain "the Central Dogma," which is to life what fusion is to the stars! This pathway, which can be summarized as the conversion of DNA (the instructions for generating living organisms) into messenger RNA (mRNA) and then into protein, is present in **all** living things and confirms we are all part of one big, beautiful family. The proteins generated within a cell determine what the cell looks like and what it can do. Most cellular proteins are enzymes, which help convert molecules (substrates) in their environment into molecules (products) they can use to survive, grow and generate offspring.

In Chapter 4, I embark on the long evolutionary road from primitive single-celled organisms to TV-watching Homo sapiens, leaving the controversial topic of how life may have started for the next chapter. I start four billion years ago when the Earth had cooled sufficiently to have liquid water (and thus allow life to exist). I indicate that competition for limited sources of energy is what drives evolution and DNA mutations are what allow evolution to happen.

I also mention that while life is incredibly resilient, existing everywhere there is water, individual species are very transient, with >99% of all the species that have ever lived now being extinct. I also describe how the advent of sexual reproduction 1.5 to 2 billion years ago led to some of the most bizarre and convoluted life cycles, especially amongst parasites such as *Toxoplasma gondii*. Approximately one third of the human population is infected with this single-celled parasite and there is some evidence it increases risk-taking behavior in its hosts to increase its own survival.

In Chapter 5, I present the currently popular theories on how life may have started. Given the background I present in Chapters 3 and 4, I then suggest that these popular theories are very unlikely to have given rise to the Central Dogma that characterizes all life today. Instead, I present a case for what is currently considered a fringe theory, interstellar spores (pansporia).

In Chapter 6, my final chapter, I share my personal thoughts on where I think we humans are heading versus where I think we should be heading, and I touch on what motivated me to write this book.

Importantly, some of what I have written will likely, one day, be proven wrong. And that's fine because that is the way science works. What I have tried to do in this book is not spend too much time on details but rather on the big picture, i.e., "not lose sight of the forest for the trees." This is not easy for me. I have spent most of my career as a biochemist, immunologist, and cancer researcher, purifying and characterizing the properties of a few proteins that no one has heard of, so I know how easy it is to get caught up in "the trees." Although I have glossed over and simplified some complex topics, I have added references for those interested in more details. You may also be surprised, but hopefully not too annoyed, to find that I have sprinkled in some humor here and there. I have done this in part to lighten some heavy subjects but also to remind you that life is over in the blink of an eye, and we Homo sapiens should not take ourselves too seriously. We are not the be-all and end-all of the universe, but rather, one life form amongst many. So just relax and don't panic if you read something that you don't understand. There will be no tests! I want you to enjoy this and if you learn something that you find interesting, even better.

OUR UNIVERSE

Chapter 1

Part 1

In this first section we are going to talk about the currently accepted model of how our universe came to be, the most important reaction in our universe, and why we, and all other living things, are truly made of stardust.

In the beginning the Universe was created. This has made a lot of people very angry and been widely regarded as a bad move

Douglas Adams

(a) In the Beginning

In the beginning…what a concept! When we, in the 21st Century, talk about the beginning of the Universe we usually start with the Big Bang Theory, which I will explain in the next paragraph. Studies to date put the Big Bang at around 13.8 billion years ago, in part, because we can't see any stars older than 13.8 billion years old. One light year is 9 trillion kms (6 trillion miles). The current size of the Universe is thought to have a width of 150 billion light years. What this means is it would take 150 billion years to go from one end of the known universe to the other if we could travel at the speed of light!! To put this ridiculous number into something we can relate to, if you could travel at the speed of light (300,000 km/sec or 180,000 miles/sec) you could go around the

Earth 7.5 times in one second!!! Incredible, right? Now imagine a universe that takes 150 billion years to go from one end to the other at this speed. The fastest we can travel in outer space today is about 61,000 km/hr (38,000 miles/hr…which is about 1/20,000th the speed of light). So, currently, it would take us 81,000 years to travel to our nearest neighboring star (Alpha Centauri, which is 4.3 light years away). Given that the sci-fi notion of wormholes in space is unlikely, we won't be visiting any neighboring stars, EVER! So, if you think the universe was all created just for we humans, it is an unnecessarily immense playground, the vast majority of which we will never, ever see.

OK, let's get started on the Big Bang theory. This theory states that the entire universe started out as a single point, where all matter and energy were concentrated into something the size of an atom (far, far smaller than the eye could see). This point then expanded at an inconceivable rate; it's thought to have occurred FASTER THAN THE SPEED OF LIGHT, which may not be possible, is not universally accepted, and may not be required to understand our current universe (1). Nonetheless, during this expansion, known as cosmic inflation, the temperature dropped rapidly from a trillion to a billion degrees centigrade (°C), which is still 100 times hotter than our sun's 10 million °C core. At this still unbelievable high temperature, positively charged subatomic particles called protons, as well as uncharged subatomic neutrons, were moving about so quickly that they began smashing into each other and fusing (becoming one). This fusion caused the release of a lot of energy in the form of high energy (i.e., short wavelength) gamma rays. Over the next 13.8 billion years these high energy gamma rays gradually lost their energy as the universe expanded and became the very low energy (i.e., long wavelength) microwave radiation that astronomers see today. In fact, one of the strongest pieces of evidence for the Big Bang theory is the presence of this cosmic microwave background (CMB) - the afterglow of the Big Bang. This background radiation fills the whole universe today and will continue to lose energy (i.e., transition to longer wavelengths) as the universe keeps expanding, becoming radio waves, which are harder to detect.

Approximately 380,000 years after the Big Bang, the expanding universe had cooled sufficiently for the simplest atoms to form. Atoms are the smallest units of an element that retain the properties of that element. They are composed of a nucleus made up of protons (positively charged subatomic particles) and neutrons (uncharged subatomic particles), surrounded by orbiting electrons (negatively charged subatomic particles). Prior to this time, atoms could not form because the expanding universe was too hot and the released light particles (gamma rays) had enough energy to knock electrons out of their orbits around nuclei. These first simple atoms were hydrogen (with only 1 proton in its nucleus, and 1 electron spinning around it), deuterium (with 1 proton and 1 neutron in its nucleus, and 1 electron spinning around it) and helium (with 2 protons and 2 neutrons in its nucleus, and 2 electrons spinning around it (see **Fig 1**). While Fig 1 shows the typical way we draw these simple atoms, electrons are actually much further away from their nucleus than depicted. As well, electrons are currently thought to exist everywhere at once like a cloud, rather than as discrete spinning electrons.

HYDROGEN	DEUTERIUM	HELIUM

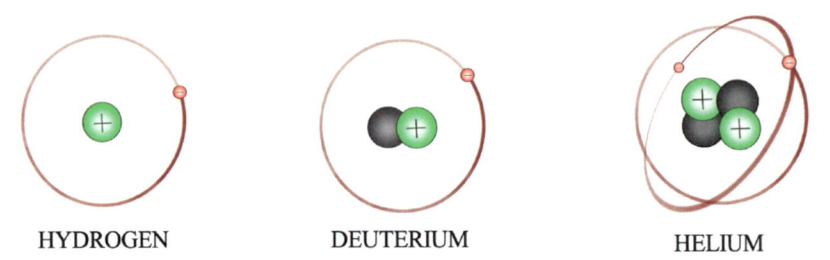

Fig 1. Simple representations of an atom of hydrogen, deuterium, and helium. Atoms are the smallest units of an element that retain the properties of that element (atom is Greek for indivisible). They typically have a nucleus in the center, composed of positively charged protons (shown in green) and uncharged neutrons (black), surrounded by circling, negatively charged electrons (red). The number of electrons is typically the same as the number of protons so that atoms are neutral (i.e., their net charge is zero). Just to drive you crazy, current thinking is that both protons and neutrons are actually made up of 3 quarks that are held together (glued together) by gluons. Electrons, on the other hand, are made up of leptons (but let's not go down this road to madness!)

A little silliness to help you remember these subatomic particles:

> A neutron walks into a bar and says, "How much for a beer?"
> The bartender says, "For you, no charge!"

> One hydrogen atom says to the other, "I think I just lost my electron!"
> The other says, "Are you sure?"
> "Yeah, I'm positive!"

OK. Time to get serious again! For a few hundred million years after the Big Bang, there were still no stars (and so this is called the Dark Age) and the universe was a dark fog made up of 75% hydrogen and 25% helium. Then, gravity caused clumps of hydrogen and helium atoms to condense to form the very first stars. This happened because anything with mass has gravity, from atoms, to stars, to you. Objects containing more mass have a stronger gravitational pull. The first randomly formed hydrogen and helium clumps had more mass than individual hydrogen and helium atoms. More mass meant more gravitational pull, so additional hydrogen and helium atoms were drawn towards these clumps. As a result, the clumps grew larger and larger, until they eventually formed the first stars… and yes, stars are just big balls of gas, composed of mainly hydrogen and helium. These first-generation stars were massive and short lived, crushed by their immense gravity and often exploding as supernova (the explosive and very luminous last stage of a massive star) (2).

(b) Fusion Provides Energy to the Universe

Now, I know this is going to hurt a bit, especially for those who have not had any science in school, but I would like to walk you through the fusion reaction that occurs in a star, since this is arguably the MOST IMPORTANT REACTION IN THE UNIVERSE! In just a few paragraphs, I will explain how this reaction is responsible for providing us with light, heat, and all the elements that make up living and non-living things… so bear with me! It is also the most abundant reaction in the universe. For example, our Sun, at its core, fuses 620 million tons of hydrogen into helium every second!!! Yes, I said every second! Without this fusion of 4 hydrogen atoms to form 1 helium atom, the universe would be dark, cold, and lifeless (and be composed of mostly hydrogen). Although a hydrogen nucleus (which is made up of 1 proton, as shown in Fig 1) is positively charged, and thus doesn't want to be near (is repelled by) another positively charged hydrogen nucleus (like-charges repel), the high pressure in a star's core causes hydrogen nuclei to get close together. The high temperature in a star's core causes these hydrogen nuclei to move so fast that when they crash into one another, they fuse (the nuclei of atoms only ever touch in stars).

As shown in **Fig 2**, when 4 protons (4 hydrogen nuclei) fuse, 2 of the 4 protons get converted into neutrons by each proton releasing a neutrino and a positron. So, remember, if you are a proton, which is positively charged, and want to become a neutron, which has no charge, you have to lose a neutrino and a positron. Neutrinos have very little mass and interact very weakly with other forms of matter. As a result, neutrinos escape the sun's core instantly and travel close to the speed of light, passing through all forms of matter without being affected (tens of thousands of neutrinos pass through your body every second!!). Positrons, on the other hand, have the same tiny mass as electrons but are positively rather than negatively charged. These positrons collide with electrons inside the sun's core. This collision destroys them both (i.e., electron-positron annihilation), completely converting them into energy (gamma rays). So, it is the release of a positron from a proton that leads to the generation of gamma rays, via the subsequent collision of the released positron with an electron!! These gamma rays then crash into many particles as they work their way to the surface of a star. In our sun, gamma rays take 10 million years to travel from the sun's core, where they are generated, to the sun's surface! Let me repeat that! Gamma rays take 10 million years to travel from the sun's core to the sun's surface! As they travel within the sun, they lose energy because of all the collisions. As a result, by the time they leave the sun's surface, they have transformed from high-energy gamma rays into lower energy visible, ultraviolet (UV), and infrared (heat) rays. They then take only 8 min +10 sec to reach us on Earth, providing us with light and heat. This is all summarized beautifully in episode 6 of the wonderful 13-episode Cosmos series put together by Carl Sagan and Neil Degrasse Tyson (3).

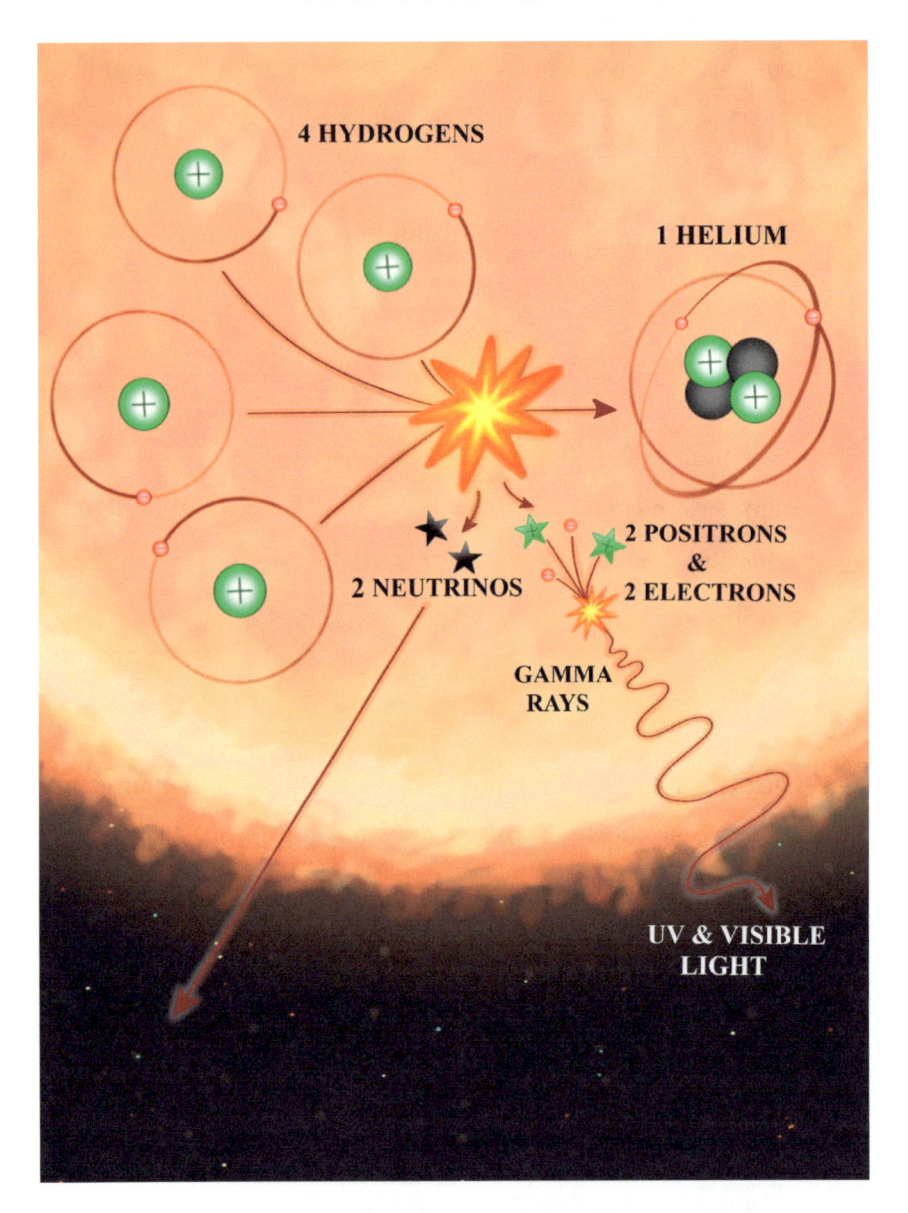

Fig 2. The fusion of 4 hydrogen atoms to become 1 helium atom in a star's core. 4 hydrogen atoms (i.e., 4 positively charged protons + 4 negatively charged electrons) fuse to form a single helium atom (made up of 2 protons and 2 neutrons in the nucleus and 2 electrons spinning around it) within the core of a star. In this process, 2 of the hydrogen protons become uncharged neutrons by losing a positron and a neutrino. The 2 generated positrons smash into 2 hydrogen

electrons, causing mutual annihilation and transformation into energy, in the form of gamma rays. By the time gamma rays travel to the surface of the sun, they have transformed into lower energy visible, ultraviolet (UV), and infrared (heat) rays. The 2 remaining hydrogen electrons become part of the helium atom.

To summarize, fusion in stars converts the lightest element, hydrogen, into the second lightest element, helium (the ash of a star), and generates energy (gamma rays) by destroying mass (i.e., positrons and electrons). So, how much energy does all this generate? Einstein's $E = mc^2$ tells us that mass and energy are inter-convertible and that the destruction of a little bit of mass (m) generates a huge amount of energy (E), since the mass must be multiplied by the constant c^2 (the speed of light squared!) to indicate how much energy is generated. To get a better feeling for this conversion of mass to energy, 1 g of mass (1/5th of a teaspoon) can generate the same level of energy that was released by the atomic bomb that fell on Nagasaki!!

Stars that are either the size of our sun or smaller can only convert hydrogen into helium, as described above. However, stars larger than our sun can generate more gravity and thus have higher pressures and temperatures in their cores. These higher pressures and temperatures make it possible for these larger stars to generate heavier elements than helium because these larger stars can overcome the higher repulsion that larger nuclei have for each other (i.e., larger nuclei repel each other more than smaller nuclei because they have more positive charge). As an aside, if there are 2 or more protons in a nucleus you need a neutron to hold them together (by acting as a mediator or buffer) because protons repel each other. In stars more massive than our sun, the generated helium can be fused to form carbon and oxygen and, in very massive stars, elements as heavy as iron (with 26 protons, 30 neutrons and 26 electrons) can be generated via fusions. Elements heavier than iron are made by supernovas. Supernovas are very bright explosions that occur when old, extremely massive stars collapse (because their hydrogen is depleted). These explosions result in the spewing forth of higher elements into the universe. **So, stars not only make life possible by providing light and heat but also all the elements that make up living things** – i.e., we are indeed made of stardust!

An important point to make here is that the first (oldest) stars to form were made up of just hydrogen and helium but when they exploded as supernovas, they released clouds of gas containing heavier elements into space. When these gas clouds condensed into new stars, these new stars contained heavier elements. As a result, we can tell how old a star is (whether it is a first, second or third generation star) by its level of heavy elements (which astronomers call "metals"). You can measure the ratio of heavy elements/hydrogen in a star via a spectrophotometric analysis of the light coming from it. Our sun is considered a third-generation star, made up of the recycled materials from earlier stars.

It is comforting to know that stars tend to be stable for millions to trillions of years because of a balance between the inward pull of gravity and the outward push of hot gases. Although very big stars tend to be less stable, as mentioned above, because their massive gravity tends to crush them, ours is just average sized, so we don't have to worry about the stability of our sun for now.

Of interest, in late 2022, scientists in California, using a powerful laser, were able for the first time to generate a fusion reaction that produced more energy than was required to trigger it (4). In the future, this discovery could lead to the harnessing of fusion as an alternate, non-polluting source of energy.

OK! I think we need a break now so put this down and let's reconvene after you have had some time to mull over what we have covered so far!

Part 2

Welcome back! You are a brave person. In Part 2, we are going to talk about galaxies, our solar system, the nature of light and gravity, an attractive alternative to the Big Bang (the Big Bounce), and the scary possibility of endless space and time (don't be scared!).

Only two things are infinite, the universe and human stupidity, and I'm not sure about the former   Attributed to **Albert Einstein**

(c) Galaxies

OK. We have covered stars and their fusion reactions that generate all the energy and elements in the universe. So how are these stars organized in the universe? The answer is they are clustered into galaxies. Currently, it is estimated that there are 100s of billions of galaxies in our universe, with each galaxy likely having a supermassive black hole at its center. Many Sci-Fi movies have explored black holes, but what are they really? When really large stars have burnt up most of their hydrogen, they no longer have the outward push of their hot gases to maintain themselves, and they collapse. As they collapse, their incredible gravity pushes their electrons right into their protons to become neutrons. If these really large neutron stars are massive enough, they turn into black holes. Black holes are so dense that neither light nor any particle can escape from them and if all the people currently living on Earth were converted into the material of a black hole, we would only constitute one teaspoon! The incredible gravitational pull of a black hole keeps its many circling stars from escaping. Each galaxy, including our own galaxy, the Milky Way, with its supermassive black hole called Sagittarius A, has, on average, 100s of billions of stars. Some galaxies are spiral-shaped, like our Milky Way, with curved arms that make it look like a pinwheel. The younger, and thus brighter, stars in these galaxies are typically in the outer arms and the older stars are in the center. Other galaxies are oval shaped and called elliptical galaxies. Still others are irregular and look like blobs (these are typically the smallest galaxies). All these galaxies are moving relative to each other, and it is predicted that in about 4 billion years our Milky Way galaxy will crash into our nearest neighbor, the Andromeda galaxy and, in time, will form a single galaxy. The good news is that because the spaces between stars are so vast, our solar system will likely not be affected by this interstellar collision. However, the bad news is that by the time these two galaxies collide, our sun will have grown into a dying, red giant and boiled away all the water on Earth (5).

(d) The Formation of Our Solar System

About 4.6 billion years ago, near the outer edge of our Milky Way, it is thought that a massive first-generation star exploded as a supernova and generated a huge cloud of slowly spinning debris (a molecular cloud), 100s of light years in size (meaning it would take 100s of years to go from one side to the other, traveling at the speed of light). This debris contained not only hydrogen and helium but many heavy elements as well, including iron, carbon, nitrogen, oxygen, aluminum, nickel, carbon, and silicon, and started to condense under the force of gravity. This contraction of the debris caused the spinning to speed up, not unlike a spinning figure skater who spins faster when their arms are brought close to their body. The center of this spinning, condensing cloud heated up and became our Sun. The surrounding disk of dust and gas may initially have coalesced into about 500 moon-sized little planets (planetesimals) that collided into one another to produce the current inner planets, Mercury, Venus, Earth, and Mars. Our Earth at this time was extremely hot because of all the collisions. Then Earth's biggest natural disaster occurred about 4.5 billion years ago. A planet half the size of Earth side-swiped our planet (the "big splat"), sending a huge amount of debris into space. This big splat may have caused both the spin and the tilt the Earth now has, resulting in our day and night cycles and our seasons, respectively (6). Over time this debris from the big splat coalesced into our moon. A silly mnemonic to help you remember the order of the planets in our solar system: **M**en **V**ery **E**arly **M**ade **J**ars **S**tand **U**p **N**early **P**erfectly (for Mercury, Venus, Earth, Mars, Jupiter, Uranus, Neptune and Pluto). If you haven't heard this before, you're welcome! Also, I know Pluto has been demoted from planetary status, but you can't end a sentence with "nearly"!

(e) Light

Now is as good a time as any to discuss the enigma of light. We used to think light was made up of waves, like sound waves. After all, like sound waves, light can be reflected, refracted (changed slightly in its direction when going from one medium into another, like from air into water) and diffracted (bent as it goes through a tiny hole in a wall). As well, light comes in different wavelengths: shorter, high-energy wavelengths, like ultraviolet (UV) and violet light (in the visible spectrum), and longer, low-energy wavelengths, like red light (in the visible spectrum) and infrared (heat), which has the least energy. All of these properties are reminiscent of sound waves, which are reflected off surfaces (which is why you hear an echo) and exist in both long (think base) and short (soprano) wavelengths.

A funny aside is in the late 1700s, the great British astronomer, William Herschel, put a quartz crystal (prism) in his window at home in England to separate the different wavelengths of sunlight streaming into his study. As a result, he had all the visible colors of the rainbow displayed on his big wooden desk. He knew that sunlight was warm and so he wanted to see which color (wavelength) was responsible for this warmth. To do this he put a thermometer within each color on his desk. As a control, he put a thermometer just outside the rainbow, right beside the red light. Strangely, he found his control thermometer showed the biggest increase in temperature! This observation is how infrared (which means below red) radiation was discovered, paving the way for microwave ovens two hundred years later!

Light also behaves, under certain circumstances, like it is made of particles. For example, when UV light hits a metal surface it knocks off electrons from the metal surface (the photoelectric effect). Another problem with thinking of light only as a wave is that sound waves are propagated by causing molecular vibrations and can only occur if there are molecules for the sound to excite (vibrate). There is no sound in outer space because there are no molecules to transmit the sound (forget those sci-fi movies with their big booms when starships blow up). Since light can travel through the vacuum of space, many scientists in the 1800s thought that there must be a mystical "ether" in space that allows light waves to be transmitted. However, the general consensus now is that light has a wave-particle duality, i.e., it is made up of massless, elementary particles called photons that travel in waves and can travel through outer space without the need to cause any mythical space particles to vibrate (7).

Let me repeat this. The smallest discrete amount (unit) of light is called a photon and is defined as a particle without any mass that is always in motion and traveling, of course, at the speed of light.

A photon checks into a hotel and the bellhop asks, "Can I help with your luggage?"

The photon replies, "Don't have any, I'm traveling light!" (I apologize! Just trying to keep things light!)

(f) Looking Back in Time

OK. So, we have covered stars, galaxies and light. Now we can talk about something really neat. When you look at the night sky filled with stars, what you are really doing is looking back in time! William Herschel, the astronomer (and very accomplished musician) mentioned above, was one of the first to propose this concept. In the late 1700s, he painstakingly built a very large telescope, which allowed him to discover the seventh planet in our solar system, Uranus (God of the sky in Greek... but also the "butt" of many bad jokes). As well, together with his sister, Caroline, William showed that distant milky patches in the night sky were actually composed of many stars (separate galaxies). In terms of going back in time, I mentioned earlier that our nearest star (after our sun), Alpha Centauri, is 4.3 light years away from us. This means that when we look at this star, the light reaching our eyes left this star 4.3 years ago, so we are actually seeing what it looked like back then. For all we know it could have blown up 2 years ago and we wouldn't know it until 2.3 years from now. As we look at stars further and further away, we look deeper and deeper into the past. The furthest stars we can see with our unaided eyes are about 4,000 light years away (which means the image of them that we see today is what they looked like 4,000 years ago!). The very furthest stars we know of are 13.8 billion light years away and may be long gone.

A logical question you might ask at this time is, how do we know how far a star is from us. This is especially complicated because some stars are brighter than others and so we might think they are closer than they are. Typically, the bigger a star, the brighter it is (i.e., more luminous, because it is burning at a higher temperature). There are two ways we currently determine the distance of a star from us. One is based on simple geometry (for stars that are close to us) and the other on the color of the star (since the color tells us the surface temperature of the star). A fun reference that explains this really well is (8).

(g) Gravity

Gravity is what enabled our current universe to form by causing molecules of hydrogen and helium to clump together and eventually form our stars. In 1687, the very reclusive and brilliant Sir Isaac Newton came up with a simple law of universal gravitation which states that gravity is a force (F) that causes any two bodies to be attracted to each other, with the force proportional ($\propto$) to the product of their masses (m) and inversely proportional to the square of the distance (r) between them. In other words, if you double the distance between any two objects, their attraction for each other is 4 times weaker.

$$F \propto \frac{M1 \times M2}{r^2}$$

Of note, mass is not quite the same thing as weight, although it is typically used as such on Earth. Mass refers to the total amount of "matter" in an object and is constant (not affected by gravity) while weight increases with increasing gravity. So, for example, your weight and mass are both 150 pounds (68 kilograms) on Earth but on the moon, which is much smaller, and so has less gravitational pull, you would weigh 24 pounds (10.9 kilograms) but still have a mass of 150 pounds. So, gravity causes mass to have weight (9).

Based on Isaac Newton's theory of universal gravitation, it has been proposed that gravity is caused by massless, uncharged particles called gravitons (10). which have never been observed but cause objects to be attracted to one another.

In 1915, Albert Einstein proposed that gravity is more accurately described by his general theory of relativity, which states that gravity is not a force, but a consequence of the curvature of spacetime (the 4th dimension) and is caused by the uneven distribution of mass, i.e., it's the warping of spacetime that causes objects' paths (including light) to curve in the presence of gravity. The usual analogy is of a planet sitting on a mattress and thereby distorting the mattress (spacetime) (11). What this theory really suggests is that time moves slower when you are close to a massive object. Because it was subsequently shown that Einstein's equations more accurately described the somewhat odd orbit of Mercury around the sun and the bending of light around massive stars than Newton's, his equations have supplanted, to some extent, Newton's law of universal gravitation. However, since Newton's equation is, for most purposes, adequate and much simpler to calculate, it is still used for working out the trajectories of current spacecraft. As an interesting aside, Albert Einstein's original general theory of relativity suggested that the universe was either expanding or contracting. However, since the general consensus in 1905 was that the universe was fixed and eternal, he put in a constant to counterbalance the effects of expansion/contraction. He later thought of this as the biggest blunder of his career (12). We now know that the universe is expanding, and this laid the groundwork for the Big Bang theory.

(h) Dark Matter and Dark Energy

Two mysterious phenomena that complicate our current understanding of the universe are dark matter and dark energy. Dark matter, as its name implies, is matter that is unseen but inferred from its gravitational pull on regular matter and may be composed of the elusive gravitons mentioned above (13). There is good evidence for its existence. According to current calculations, normal matter, made up of atoms, may account for only 5% of the universe while dark matter may comprise 25% of the universe. Dark energy, on the other hand, which may make up 70% of the matter + energy of the universe, was originally invoked to explain the recent observation that every galaxy is actually accelerating away from every other galaxy. This idea that the galaxies are accelerating away from each other is disturbing since, before this discovery, the general consensus was that the expansion of the universe would slow over time because of gravity. To explain what could be causing this acceleration, scientists came up with a repulsive force that counteracts gravity, i.e., dark energy. However, not all cosmologists accept this acceleration of the universe and its accompanying need for dark energy (14). As well, even amongst the majority of cosmologists that accept dark energy and the current acceleration of galaxies away from each other, a 2022 paper suggests that dark energy may be decaying/weakening, and the acceleration of the universe may end in the next 65 million years. The end of acceleration would then be followed by a halt in expansion and the beginning of a contraction (15). We will have to wait and see how this plays out!

(i) Space and Time

What I want to talk about now is not our terrifyingly dangerous, beautiful and enormous universe, as interesting as it is, but the empty space beyond it. So, imagine for a second our expanding universe, with its billions of galaxies, exploding out into the vacuum of space (see **Fig 3A**). See the empty space that surrounds our universe. Now imagine this surrounding empty space going on forever. To make this "forever" easier to imagine, envision a box enclosing our current universe as shown in **Fig 3B.** When we travel out in space and reach the inside wall of this box, we can't help but wonder what is on the other side. There must be something there, even if it is only empty space. Based on this logic you either have to keep making the box bigger or accept that space must go on FOREVER, as shown in **Fig 3C.** I can't take credit for this crazy concept. The ancient Roman poet/philosopher, Lucretius, who lived before Christ, proposed a limitless universe and there is evidence that some ancient Greek philosophers, long before Lucretius, did so as well. OK, before we continue, I have to say that while I can intellectually accept space going on forever, since there is really no other option, I can't really grasp the concept. Our brains evolved in a finite world and the notion is just too foreign. As Simone de Beauvoir said *"I am incapable of conceiving infinity, and yet I do not accept finity"*.

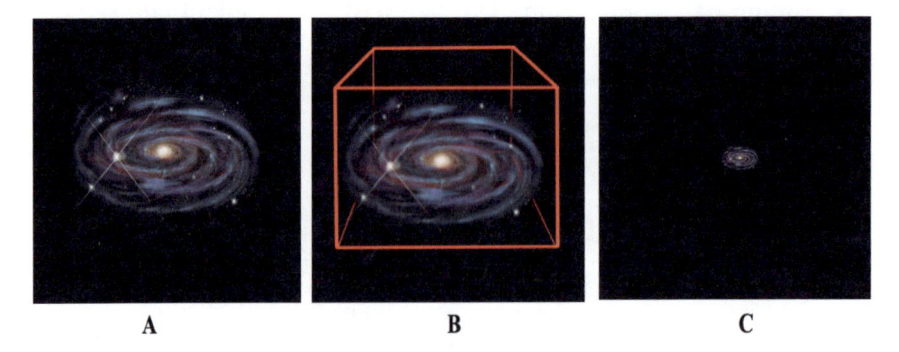

A   B   C

Fig 3. Different views of our expanding universe.

But OK, if you accept this strangely disquieting idea of infinite space, our "little" universe is in an awfully big playground, and thus it seems unlikely that it is the only one. So, there is a possibility that other universes exist (defining a universe here as originating from a Big Bang).

Let's accept for a moment that space is infinite and get even more controversial. Let's talk about time. One could imagine in the future that our Universe continues to cool – i.e., today's universe has an average temperature of only 2.725 Kelvin (about -270°C, which is very close to 0 Kelvin, where all thermal motion ceases). If this cooling continues, the most probable fate for our universe is a total loss of energy and, at this point in time, all motion stops. Once motion stops, there are 2 possibilities: a Big Crunch or Dissipation.

In possibility #1, gravity starts a contraction of our universe - a Big Crunch - and our galaxies start accelerating back toward each other, perhaps towards the site of the original Big Bang. The clustering of galaxies creates an incredibly massive gravitational pull that crushes the galaxies together. This crush generates so much energy that a new Big Bang ensues, in which the universe starts expanding again. In other words, the universe cycles indefinitely between expansion and contraction (16). Again, this is not a new idea. For example, the Hindu Rigveda, written in India around the 15th - 12th Century B.C., describes a cyclical or oscillating universe in which a "cosmic egg" or Brahmanda, containing the whole universe, expands out of a single concentrated point called a Bindu before subsequently collapsing again. Adherents of this expansion and contraction cycle prefer the term Big Bounce rather than Big Bang. The reason this idea is important in cosmology is that we invoked "inflation" (i.e., the faster than light expansion after the Big Bang) to explain some of the properties of our current universe (i.e., its even temperature and homogenous appearance), but if there was a contraction before the Big Bang, we don't have to invoke inflation to get this appearance (1). Going further with this possibility, one can imagine looking back in time, seeing our own Big Bang occurring 13.8 billion years ago and then further back to the contraction that led to our Big Bang and then further back to countless Big Bangs (Bounces) and contractions before that.

Alternatively, in possibility #2, the repulsive force of dark energy (which may or may not exist and may or may not be weakening) causes our accelerating universe to keep expanding until it totally dissipates into the vastness of space. This theory, which is referred to as the Big Chill or Big Freeze, is currently in vogue amongst most cosmologists. However, possibility #2 doesn't explain how the Big Bang started in the first place, a problem that is removed with the Big Bounce. Regardless of what happens to our universe, if there are other universes (i.e., we are in a multiverse), other Big Bangs will continue to be generated and die. This continuation is consistent with both space and time being infinite.

However, there may be more than these two possible scenarios for the end of our universe. In a very entertaining book called "The End of Everything," the astrophysicist, Katie Mack describes 5 possible ways our universe may end, and I encourage those interested to read it (5). Is the idea that time and space are infinite somewhat disturbing and beyond our grasp? Definitely!!

Nonetheless, contrary to the title of this chapter, I would propose that there really is no beginning… and no end.

To recap, we think our universe came into existence about 13.8 billion years ago, based in part on the fact that we can't see any stars further than 13.8 billion light years away. Our universe likely started as a Big Bang or Big Bounce, based on the background cosmic radiation that we see today. The most important reaction in the universe is the fusion of hydrogen into helium. This reaction converts a very small amount of mass (positrons and electrons) into energy. Large stars and supernovas also spew out heavier elements. These elements, together with the light that all stars generate, make life, as we know it, possible. It is likely that space is infinite and that there are other universes out there, at different stages of expansion and either contraction or dissipation. It is also possible that time is endless, with no beginning and no end.

EARTH - A GOLDILOCKS PLANET

Chapter 2

In this chapter we are going to talk about what kind of a planet you need in order to allow life to appear and thrive.

Good planets are hard to find... don't blow it **Anonymous**

(a) Setting the Stage for Life

As mentioned in the last chapter, stars not only provide the energy (light and heat) needed to sustain life, but also the elements used to make the molecules present in all living things (at least all living things on this planet) and all non-living things as well. While recent data suggest that most stars have a number of planets around them, most of these planets are unsuitable for life. So, before we talk about life itself, let's talk about what a life-sustaining planet looks like.

(b) The Importance of an Atmosphere

First off, we need a planet big enough to hold onto an atmosphere. If the planet is too small, it won't have a strong enough gravitational pull to hold onto any gas molecules that typically make up an atmosphere. While Earth is big enough to hold onto an atmosphere, it is still relatively small (with a relatively weak gravity) and so its atmosphere is made up of heavy **molecules** like nitrogen (N_2), oxygen (O_2), carbon dioxide (CO_2), and water vapor

(H_2O). To explain why Earth's atmosphere has only heavy molecules, I have to backtrack for a moment to Sir Isaac Newton's law of universal gravitation, mentioned in Chapter 1. Specifically, I would like you to remember that the gravitational pull between any two objects increases if **either** or both of their masses increase, so the Earth has a bigger gravitational pull on a more massive gas molecule than a less massive gas molecule. To calculate the mass of any atom you simply add up the number of protons and neutrons (electrons are too small to contribute significantly). Nitrogen, which makes up 78% of our current atmosphere, is relatively heavy. The nitrogen atom, for example has 7 protons + 7 neutrons, so has a relative mass of 14. Complicating things slightly, individual nitrogen atoms, as well as oxygen and hydrogen atoms, are unstable and form stable dimers with themselves (form homodimers). So, nitrogen molecules (N_2) in our atmosphere are made up to two nitrogen atoms and so the molecular weight of N_2 is 14 x 2 = 28. Similarly, oxygen atoms (O) (with 8 protons and 8 neutrons) = 16 x 2 = 32 for each oxygen molecule (O_2). The hydrogen molecule (H_2), on the other hand (with only 1 proton per atom) = 1 x 1 = 2 is the lightest molecule in existence. So, the Earth is able to hold onto nitrogen and oxygen more strongly than to hydrogen.

As well, heavy gas molecules move more slowly and are therefore easier to hold onto by a planet with weak gravity. The lighter molecules like hydrogen and helium will escape or exist only at the upper-most reaches of a small planet's atmosphere. Larger planets, on the other hand, have thicker atmospheres because their larger gravitational pull can not only hold more gas molecules but lighter ones as well. As you might expect, gravity and temperature work in opposite directions, with a higher temperature causing an atmosphere to boil away and a higher gravity retaining an atmosphere.

So why do we need an atmosphere for life, you ask? Because it supplies atmospheric pressure. Specifically, the air above you is pushing down on your head and shoulders with a weight of 14.7 pounds/square inch (14.2 kg/cm^2), if you are at sea level. And why is this important? Because this air pressure is needed to keep water from boiling away into space. And this is important, in turn, BECAUSE WATER IS ESSENTIAL FOR LIFE (that is, life as we know it. I'm going to stop saying this now but always keep this in mind!). To maintain an atmosphere conducive to life, we also need a "temperate climate" (i.e., a "Goldilocks" temperature between the freezing and boiling points of water) because if it's too hot, the gas molecules will speed up and escape the pull of gravity and if it is too cold, life cannot flourish in ice.

Our atmosphere, in relation to our planet, is very thin (like the skin on an apple) and fades away at about 300 miles (480 km) above the Earth's surface. Scientists have divided our atmosphere into 4 or 5 layers, based on temperature. Starting from the Earth, there is the troposphere, which is the warmest layer, reaching up to 6.2 miles (10 km) above sea level. This layer is where most of the clouds are and where the commercial jets fly. Above the troposphere are the stratosphere, the mesosphere, and the thermosphere (the furthest and coldest). Some people add on a 5th layer, called the exosphere, which fizzles off into outer space. Interestingly, in the late 1700s when people started going up in hot air balloons **within** the troposphere, they expected it to get warmer as they got higher, since they were getting closer to the sun, but it got colder instead. The reason it got colder is that the sun is much further away from Earth than they thought at the time, and they were only getting closer by a negligibly small amount. Importantly, it becomes colder as you go further away from the Earth's surface because the air gets "thinner" (i.e., there are fewer atmospheric molecules for the sun's rays to excite and so there are less of them bumping into each other to generate heat).

Even though our whole atmosphere forms only a very thin layer over the Earth, it is extremely important, not only because it keeps water from boiling away into space, as already mentioned, but because it has, within its stratosphere, a thin layer of ozone. An ozone molecule is composed of 3 oxygen atoms (O_3), unlike the standard oxygen molecule we breathe with its 2 oxygen atoms (O_2), and it absorbs the most damaging ultraviolet (UV) light from the sun (UV-C, which has the most energy within the UV spectrum). UV-A, a subcategory of UV light with the least energy and the longest wavelength, and some of the intermediate UV-B, are not absorbed by the ozone layer and reach Earth to give us sunburns. You may remember when chlorofluorocarbons (CFCs) were banned because they were depleting the ozone layer. This ban is one of our few successes at saving the planet! With some luck we may have more such successes! What I will talk about in Chapter 4 is how life itself has changed the makeup of the atmosphere.

Importantly, an atmosphere also limits the energy lost to space from the planet's surface, via the greenhouse effect. Given the current global warming crisis, I think I should spend a little time talking about this greenhouse effect.

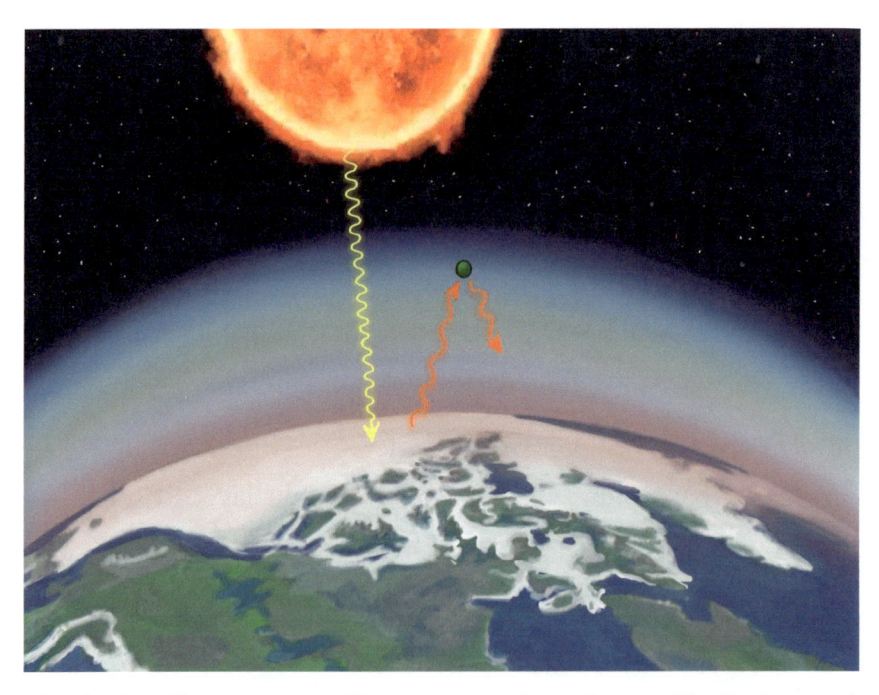

Fig 4. The Greenhouse effect. Energy from the sun (yellow arrow) warms the Earth's surface. Some of this energy is reflected back into the atmosphere with reduced energy (orange arrow) where it either escapes back into space or is deflected by greenhouse gas molecules (green ball = water vapor, carbon dioxide, and/or methane) back to the Earth to warm it.

Greenhouse gases absorb light and heat coming from the Earth's surface and reflect it back down to the surface. These gases, such as water vapor, ozone, carbon dioxide, and methane are both naturally occurring and man-made. You can infer from **Fig 4** that the more greenhouse gas molecules there are in our atmosphere, the more light and heat will be reflected back down to Earth to warm it up. While greenhouse gases keep the Earth from becoming a snowball, we have increased it to dangerous levels by burning fossil fuels (which generate excessive amounts of carbon dioxide). We have also increased it to dangerous levels by populating the planet with cows (that fart methane, a far more potent greenhouse gas than carbon dioxide) and clear-cutting forests (which, if not cut down, absorb carbon dioxide and thus reduce greenhouse gases). This has resulted in runaway global warming (17). The problem is that global warming generates a positive feedback loop that accelerates global warming. For example, global warming melts our ice caps, which is bad because ice is very good at reflecting heat back into space. As well, as the air warms up it can hold more water vapor, reflecting more heat down to the planet. More water vapor in the air also means more heavy rains and hurricanes in the future. Apart from moving away from fossil fuels, our real problem is our burgeoning human population, which is overwhelming the resources of our planet. Although improvements in agriculture have delayed massive starvation, the predicted increase in our population from 8 billion today (end of 2022) to 10 billion by 2050, coupled with global warming and its accompanying desertification (drying out of the land), will end up causing famines, serious disease outbreaks and wars (for shrinking resources). Interestingly, Thomas Malthus predicted this way back in 1798. So, we must exercise birth control now!! I am not proposing this be done through governmental edicts, but rather through education and concern for our species and our planet.

(c) Air Pressure and Wind

The gas molecules in our atmosphere have weight. As already mentioned, the weight of the air pressing down on you at sea level is 14.7 pounds/square inch (which is equivalent to 760 millimeters of mercury in a barometer). However, if air is heated (say at the equator), the atmospheric gas molecules gain more energy and expand (the gas molecules get further away from each other), making them less dense. As their density decreases, they rise up, like heating air in a hot air balloon. This rise in gas molecules creates a low-pressure zone (a relative vacuum) underneath the rising air. This causes cold air from cooler parts of the world, like over the oceans (high pressure zones), to rush in, creating wind. The greater the difference in air pressure (also called barometric pressure) between two zones, the stronger the winds. However, where this wind comes from is also determined in part by the spinning of the Earth, which deflects the wind direction to some degree.

(d) A Closer Look at our Planet

For life, in addition to an atmosphere, it is best to have a youngish planet that still has a molten metal core. A molten metal core is generated during a planet's birth and cools as it ages. This molten metal core revolves as the planet revolves, generating a protective magnetic field that combats cosmic ray-induced damage to living things. (Don't worry! I will elaborate on what this last sentence means in the next paragraph!).

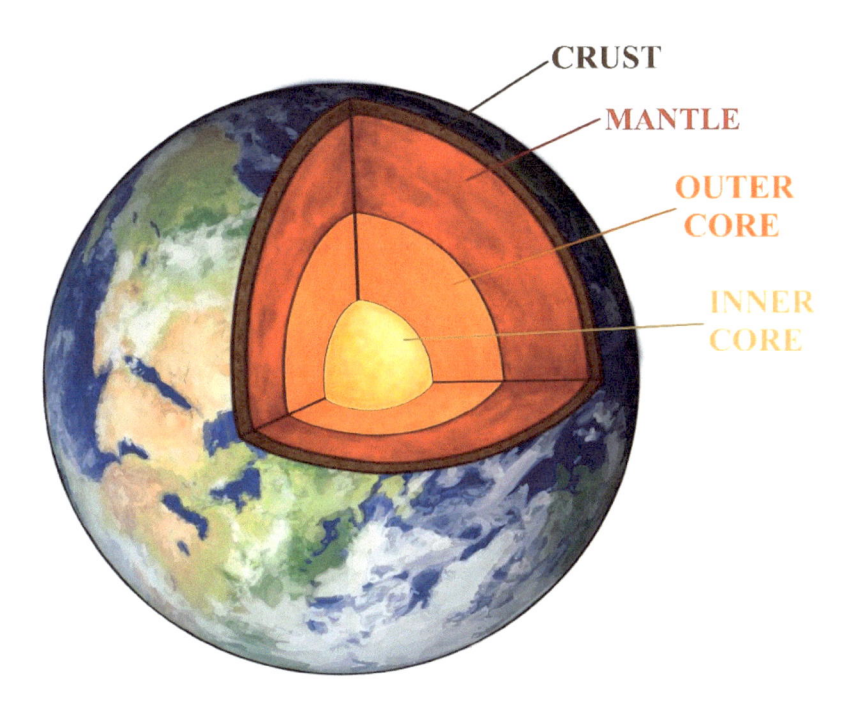

Fig 5. A look inside our Earth. The innermost core (yellow in the figure) is actually thought to be a solid ball of iron about the size of our moon, spinning within a liquid outer core. The outer core (orange) is mostly made up of iron and nickel and churns, generating a magnetic field that deflects cosmic rays from lethally mutating

life. Above this core is the mantle (red), which is semi-solid and makes up the majority of Earth's total volume (84%). On top of the mantle, is a thin crust (brown) (only about 5 - 25 miles deep). It is thought that as the Earth cooled, the heavier materials like iron and nickel sank to the center while the lighter ones (basalts and granites) rose to become this crust. On the landmasses of Earth, the very topmost layer of the crust is soil (also called dirt or earth). Soil is an extremely thin layer on top of the bedrock that makes up the vast majority of the crust. Soil is generated slowly, each inch (2.54 cm) taking between 800 to 1000 years to form, by the weathering of the bedrock (18).

(e) The Earth's Magnetic Field and Cosmic Rays

Cosmic rays are composed of extremely high-energy particles. They are generated by our sun during solar flares (also called solar wind) and by exploding stars outside of our solar system. They travel almost at the speed of light and are mainly composed of positively charged hydrogen and helium nuclei (i.e., only nuclei, because their electrons have been stripped off). These particles would cause extensive damage to our DNA (our hereditary material, which I cover in the next chapter) if they reached us but, fortunately, Earth's magnetic field (magnetosphere), generated by its churning outer core (and perhaps solid inner core as well) deflects them into space. This prevents cosmic rays from both directly damaging our DNA and from stripping away the ozone layer that protects us from UV radiation-induced DNA damage. Some of these cosmic rays also get deflected to the north and south poles where they react with molecules high in the atmosphere to generate beautiful auroras (borealis around the north and australis around the south pole).

Interestingly, while all the planets in our solar system probably arose about the same time, i.e., 4.5-4.6 billion years ago, the fact that Mars is only about half the size of Earth has resulted in its core cooling more quickly. As a result, it does not have a magnetosphere today and so its ozone layer has been stripped away and made life untenable, at least at its surface.

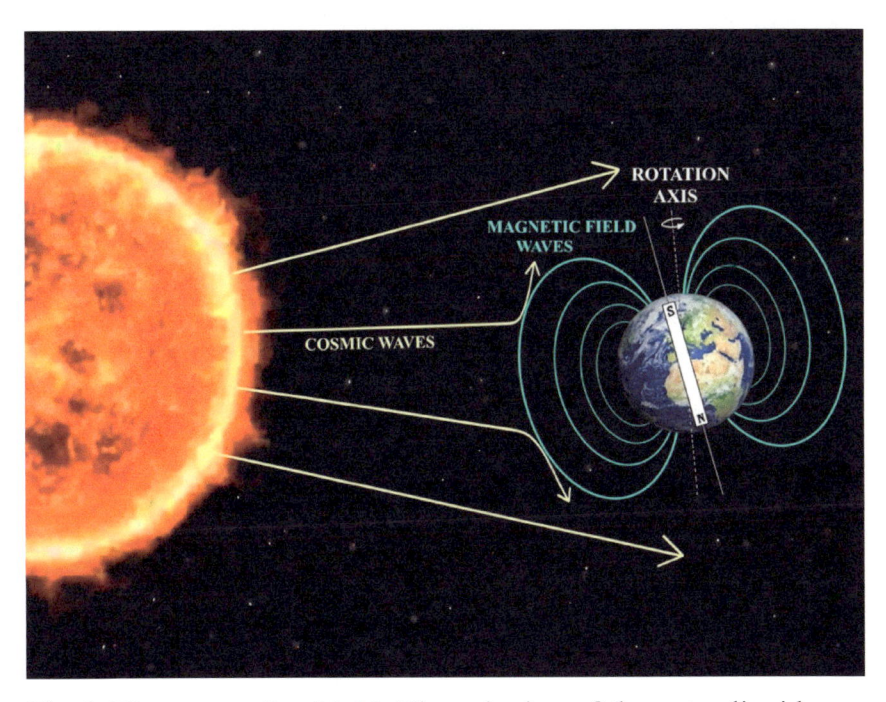

Fig 6. The magnetic shield. The spinning of the outer liquid core of our Earth generates a magnetic field that extends far beyond our atmosphere and protects us from DNA-damaging cosmic rays that come from our sun and from exploding stars in the universe.

(f) Tectonic Plates, Volcanoes, Earthquakes and Mountain Ranges

OK. We are almost at the end of this chapter on our goldilocks planet. The last thing I want to mention is our planet's tectonic plates. These are huge slabs of the Earth's crust, which fit together like pieces of a jigsaw puzzle and move because they are floating on top of the semi-solid mantle. The semi-solid mantle underneath is moving because of hot rocks rising, giving off heat, and then falling back toward the core. This movement jostles the 8-12 big and 20 or so small tectonic plates floating on the Earth's surface, causing the plates to move towards or away from each other at approximately the same rate as our fingernails grow (about 10 cm/year). Volcanoes and earthquakes occur at the boundaries of these plates when they move. So, if a volcano is erupting in your neighbourhood and you feel inclined to sacrifice a virgin, just remember it is only tectonic plates moving.

Importantly, many mountains were formed, and continue to be formed, as a result of Earth's tectonic plates smashing together, causing huge slabs of rock to be pushed up into the air (19). For example, India used to be an island, but tectonic plate movement caused it to slowly crash into the continent of Asia and form not only the Himalayas, which contain the highest mountains in the world (including Mount Everest), but the vast Tibetan plateau behind them. This process hasn't stopped, and The Himalayas are continuing to rise by an average of 2 cm each year (20).

Also of note, there is something called the mid-Atlantic ridge which is the meeting place of tectonic plates at the bottom of the Atlantic Ocean. This ridge goes on for many miles between the Eurasian and the North American Plates (at the bottom of the North Atlantic Ocean) and also between the African and South American Plates (at the bottom of the South Atlantic Ocean) (see **Fig 7**). It is at this ridge where lava spews forth constantly, generating new ocean crust on either side of it, one moving towards North America and the other towards Europe. This expansion of the seafloor will eventually result in the Atlantic being bigger than the Pacific Ocean (but not in your lifetime).

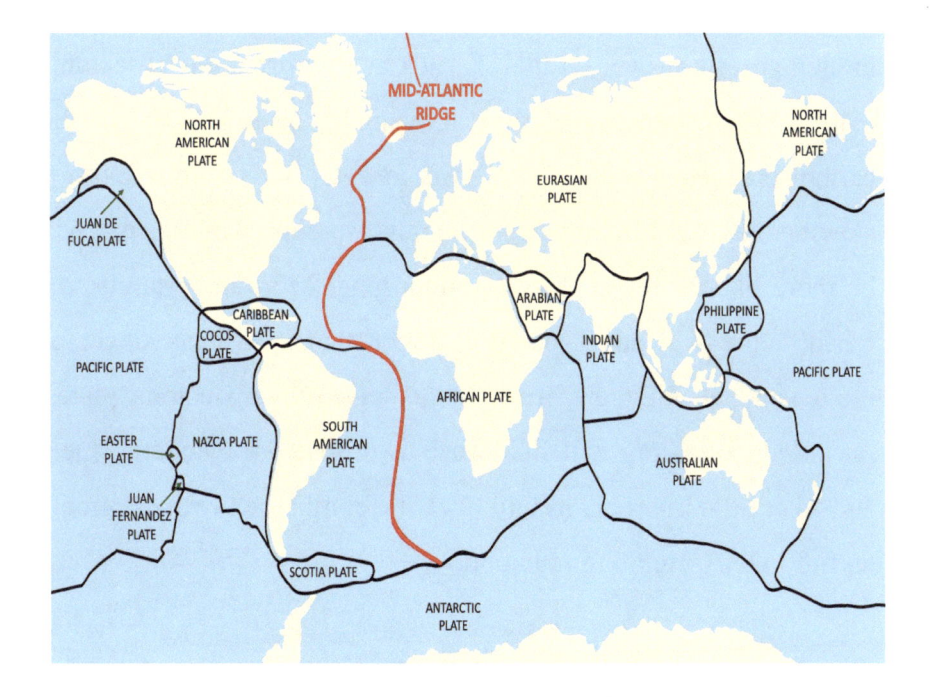

Fig 7. Earth's tectonic plates. It is likely that the landmasses of Earth have drifted apart and come back together as a single super-continent several times throughout the history of this planet.

Approximately 200 to 300 million years ago, tectonic forces broke apart the most recent super-continent known as Pangea, eventually forming the continents we have today. Shown are the boundaries of some of Earth's tectonic plates (in black), with the line in red indicating the mid-Atlantic ridge mentioned in the text.

To recap this chapter, a planet is conducive to life if it has an atmosphere, because an atmosphere prevents water, which is crucial for life, from boiling away into space. It is also important to have a surface temperature between the freezing and boiling points of water, at least part of the time, so that life can thrive. Furthermore, it helps if a planet has a molten core of spinning metal capable of generating a magnetic field so that life can be protected from DNA-damaging cosmic and high-energy UV rays. The downside of a molten core, however, is that it leads to the formation of tectonic plates, which, because of their movement, cause volcanoes and earthquakes. However, nothing lasts forever; our spinning Earth is slowing down. A billion years from now (so don't panic), a day will be more like 28 rather than 24 hours long. More importantly, the Earth's core will have cooled so much that it will no longer be molten and we will therefore no longer have moving tectonic plates. A cooled, solid core will also cause us to lose our magnetic shield and what little life remains will likely be bombarded with extremely harmful DNA-mutating radiation.

Our Universe and How We Got Here

LIFE

Chapter 3

Part 1

In this section we are going to talk about how all life is made up of cells, the similarities and differences amongst the different cell types that exist today, and proteins, the workhorses of all living things.

Life is full of misery, loneliness, and suffering – and it's over much too soon   **Woody Allen**

(a) Different Cell Types

So, what is life? First of all, all living things are made up of cells. Don't think of prison cells here, although both jail cells and biological cells (the units of life) are called cells because they are reminiscent of the little rooms that monks used to live in (from the Latin, cella). In the late 1600s when Sir Robert Hooke looked down his homemade light microscope at a thin slice of cork (tree bark) he saw "little rooms" whose walls were made of the remaining cell walls of once living plant cells. Simplistically, you can think of a living cell as a tiny bubble that is just below our ability to see with the naked eye. Importantly, most living things today are made up of just one cell (unicellular organisms). For example, one cup of soil contains more unicellular organisms (i.e., bacteria, etc.) than there are people on Earth! We humans, on the other hand, are composed of about 30 to 40 trillion cells!!! And all of these cells in our body work together in harmony (most of the time) to keep us alive and running smoothly.

The inside of all cells is composed mostly of water with some salt dissolved in it (at a concentration slightly less than our oceans) and lots of proteins (which I will get to shortly). **Fig 8** compares the cells of the simplest single-celled organisms like bacteria (which are prokaryotes, meaning they do not have a real nucleus) with more advanced organisms (eukaryotes, with a real nucleus), some being single-celled (protists) and some being multicellular, like plants and animals. A nucleus is a separate "little room" inside a eukaryotic cell that houses our genetic material (DNA, that I will talk more about later). As mentioned above, all of these different prokaryotic and eukaryotic cells are too small to see with the naked eye (except for some very rare giant cells). I have also included a virus in Fig 8 to give you an idea of its relative size (about $1/100^{th}$ the size of most bacteria and can only be seen with an electron microscope) but I do not classify these parasites as living entities. Viruses are not composed of cells but rather are simply made up of genetic material (deoxyribonucleic acid (DNA) or ribonucleic acid (RNA)) enclosed in a protein capsule. They have no biological activity on their own and need to invade cells and take over the cell's machinery in order to reproduce. The variety of cells that viruses can invade is called their "host range" and is determined by the attraction of a viruses' outer protein capsule to the proteins on the surface of cells. If the affinity is high, the virus binds to the host cell, enters the cell, and then takes control of the host's machinery to make 1000s of copies of itself. The progeny (children) of the virus then either explode out of the host cell, killing it, or "bud" out (which does not kill the host cell). If the progeny bud out, they get covered in the host cell membrane as they exit, which protects them, to some extent, from being detected by the immune system of the host.

Of note, not all viruses attack animal or plant cells. In fact, most viruses attack only bacteria, and these viruses are called bacteriophages (bacteria eaters) or phages for short. In our oceans there are 10^{30} phages (that's 30 zeros after the 10!!!, more than the number of stars in our universe), and they kill about 40% of the bacteria in our oceans every day! There are some researchers who are trying to use phages to kill antibiotic-resistant bacteria since, as just mentioned, phages only kill specific types of bacteria and do not attack/harm human cells. To combat these phages, the bacteria have evolved several defense mechanisms, including killing themselves by activating enzymes that chew up their own DNA and thus prevent the phage from reproducing (they selflessly take one for the team). A more sophisticated mechanism some bacteria have developed, if they survive an attack by a phage, involves storing little pieces of the invading viral genetic material so that they "remember" the attacker and can chop up its DNA the next time it invades. This latter mechanism, called CRISPR, was only recently discovered and is now being used by scientists to more easily edit our own DNA to correct hereditary mutations (21).

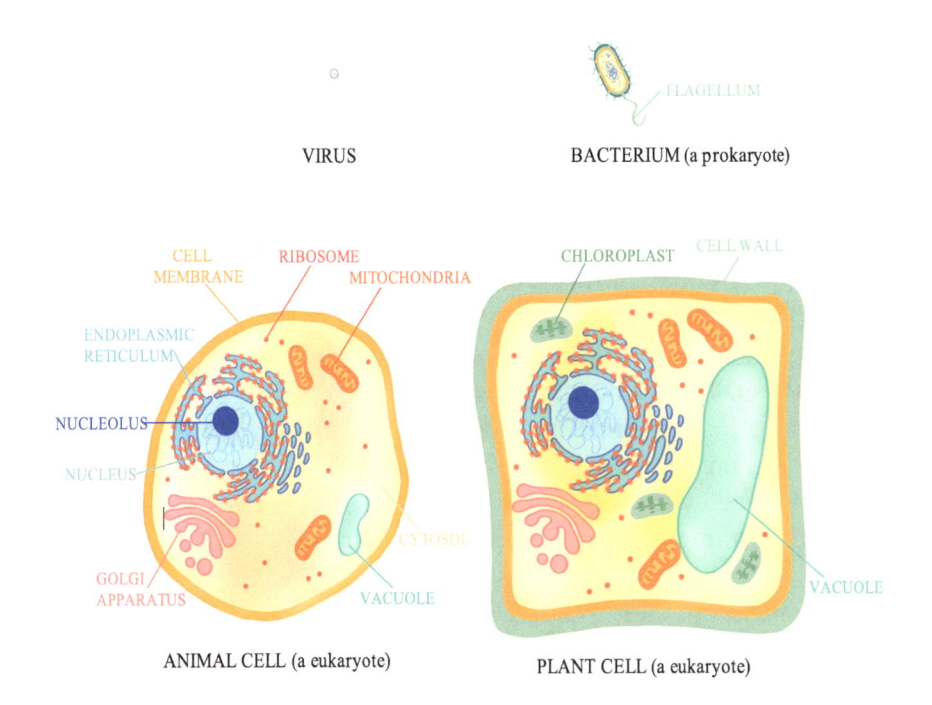

Fig 8. The Relative Size and Composition of a Virus versus a Bacterial, Animal and Plant Cell. Viruses, composed of only DNA or RNA (genetic material) encapsulated in protein coats, vary in size but are all far smaller than any bacteria. Bacteria are called prokaryotes ("before a real nucleus", with their DNA floating freely in the cytosol). They are about $1/10^{th}$ the size ($1/40^{th}$ the volume) of animal or plant cells. They often have **flagella** (whip-like tails) to allow them to move in water. Eukaryotic cells (cells with a true nucleus), in contrast, have many organelles ("little rooms") including a **nucleus** (shown in light blue) containing their DNA and a **nucleolus** (shown in dark blue) within the nucleus where ribosomes are assembled. Other organelles within both animal and plant cells include **mitochondria**, shown in orange, where most of the cell's energy is generated, and **vacuoles** (which are much bigger in plant than animal cells, are filled mainly with water, and play an important role in maintaining turgor pressure within plant cells).

Turgor pressure pushes a cell's cell membrane (also called plasma membrane) against its cell wall if the vacuole is full of water, giving the plant rigidity. This is why your plants don't wilt if you water them. An organelle specific to plant cells is the **chloroplast** (shown in green) which uses its light-capturing pigment, chlorophyll, to convert carbon dioxide and water into sugar and oxygen). All these organelles are floating in fluid (**cytosol**) that is encapsulated within the **cell membrane** (shown in orange around bacterial, animal and plant cells). The organelles + the cytosol = the cytoplasm. Both prokaryotes and eukaryotes have thousands of **ribosomes** (shown as little red dots) which make proteins. Some of these ribosomes are free floating (and make proteins that stay in the cytosol) while others are attached to intracellular tunnels (the **endoplasmic reticulum** (outlined in blue) and the **golgi apparatus** (shown in pink) and make proteins that are either secreted from the cell or incorporated into cell membranes. Bacterial and plant cells also have an outer protective **cell wall** (shown in green). Not shown is Archaea (Greek for ancient things), a group of bacteria-like single celled prokaryotes discovered in the late 1970s that do not like oxygen (i.e., are anaerobes). Although Archaea look just like bacteria when viewed through a light microscope, they have significant biochemical differences that make them capable of living in very extreme conditions and they may be the earliest form of life on Earth (22).

Now the first thing you'll notice when comparing the cells in **Fig 8** is that eukaryotic cells are much bigger than prokaryotic cells (sometimes 100 times the volume) and have lots of intracellular compartments (organelles). So why do bigger cells have all these compartments, which are like little separate "rooms" inside them? The reason is that for organisms to survive they must be able to react quickly to changes in their environment. As a single-celled organism gets bigger (which gives it **a tremendous survival advantage!!**) it is harder for biochemical reactions to occur quickly inside it because the "right" molecules, i.e., enzymes and their substrates (which I will discuss shortly) have a more difficult time finding each other. To get around this challenge, larger cells have concentrated specific enzymes and their substrates into small "rooms" or organelles so that reactions can occur faster, dramatically increasing their chances for survival. There are several different types of organelles, each one being responsible for a different function, and thus they have different enzymes and substrates concentrated within them. In fact, eukaryotic cells are called eukaryotes because they have what is called a "true nucleus", i.e., a separate "room" that houses their genetic material (their DNA). The smaller, more primitive cells like bacteria and Archaea (see the **Fig 8** legend), which are devoid of "rooms", are called prokaryotes, which means their DNA is not sequestered within a nucleus. In addition, really large plant and animal cells often get their cytoplasm (i.e., their intracellular fluid (cytosol) + intracellular organelles) to "stream" around their nuclei to speed up reactions.

A feature shared by bacteria, Archaea, and plant cells is an outer cell wall (**Fig 8**). This cell wall not only protects cells from microbial attack and physical damage, but also gives the cell rigidity so that a plant, for example, can stand upright without bones (i.e., without an endoskeleton). A cell wall also prevents bacteria, Archaea, and plant cells from exploding if too much water enters or from dehydrating if there is too little water outside of them. In plant cells, this outer cell wall is composed largely of cellulose, the most abundant organic (i.e., carbon and hydrogen-containing) polymer on Earth. In addition, some bacteria have a capsule on the outside of their cell wall. This outer capsule acts as an additional protective barrier from attack and dehydration and, importantly, enables these bacteria to attach to surfaces in their environment. Animal cells (unlike bacteria, Archaea, and plants) don't have a cell wall (or an outer capsule) and, instead, have cholesterol in their cell membrane. This cholesterol increases the fluidity of animal cells, enabling them to change shape and move.

(b) Proteins, the Work Horses of Life

Now if I were to ask you what the difference is between living and non-living things, you might say that living things are much more complex, can reproduce, move or grow, and all of these things are true. But at a molecular level what makes living things distinct from non-living things are enzymes. Simplistically, enzymes convert molecules (substrates) that a cell gets from its environment into molecules (products) that it can use for its growth and survival. The vast majority of enzymes are proteins and when I talked earlier about the "right" molecules finding/bumping into each other, I meant substrates bumping into enzymes so that the enzymes can turn them into products. So, how does an enzyme do this? First of all, the affinity (attraction) of one molecule for another, whether we are talking about a virus binding to a protein on the surface of a cell that it is invading (a host cell) or an enzyme binding to a substrate, is determined by complementary shapes and charges (**Fig 9**).

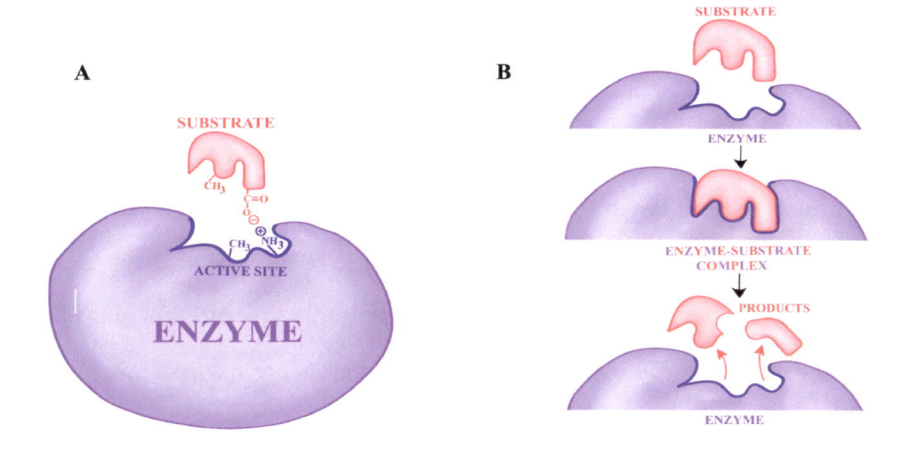

Fig 9. How an enzyme works. (A) A substrate binds to a small region of the enzyme called the "active site" because of both complementary shapes (like a key fitting into a lock) and affinities between groups within the substrate and the enzyme's active site. For example, as shown in (A), a negatively charged group on the substrate can be attracted to a positively charged group within the active site of the enzyme and parts of the substrate that are hydrophobic (hate water, i.e., typically groups made up of just carbon and hydrogen, e.g., CH_3) will interact with similar hydrophobic groups in the active site. **(B) The Induced Fit model.** When a substrate binds to an enzyme, the shape of the enzyme changes slightly to improve the binding even more. This puts pressure on a bond in the substrate that makes it easier to break, resulting in the formation of a product or products. This is the currently accepted "induced fit" model of how an enzyme converts a substrate into a product(s).

OK. So, what does a protein look like? Imagine a long piece of string with differently colored beads on it (**Fig 10**). Each colored bead represents one of 20 different building blocks (amino acids) that make up proteins. Since each bead on the string can be one of 20 different amino acids, there are almost limitless possibilities for a protein, given that a typical protein is about 300 amino acids long (but some proteins can be as short as 50 and others as long as 2000 amino acids). Each amino acid is held to its neighboring amino acids by a strong (covalent) bond called a peptide bond. Some amino acids love water (are hydrophilic) and so are happy to interact with the watery medium they are in. Others hate water (are hydrophobic) and so would rather associate with each other than the watery medium they are in. As a result, the string typically folds up into a ball with the hydrophobic amino acids hidden in the center of the ball and the hydrophilic amino acids on the outside of the ball, interacting with water. Because each protein takes on a specific shape based on its sequence of amino acids, they all look somewhat different and thus can carry out different functions. To denote proteins that are enzymes (convert substrates into products) their names end with the letters "ase". For example, lactase is an enzyme that breaks the milk sugar, lactose, into glucose and galactose (sugars, by the way, end in the letters "ose"). All enzymes have active sites that bind substrates. These active sites are usually little clefts or depressions within the three-dimensional shape of the protein.

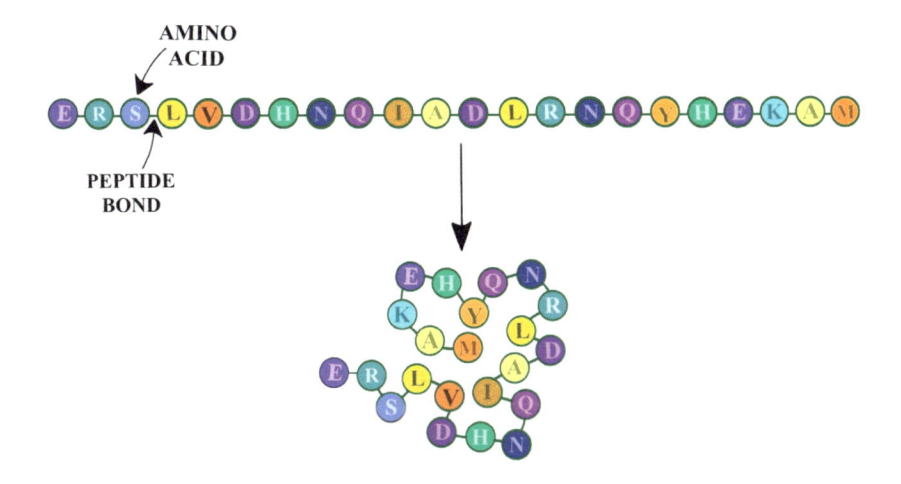

Fig 10. The structure of proteins. Proteins are simple "strings" of amino acids, held together by peptide bonds. Shown are the single letter designations for some of these amino acids. Proteins are assembled on ribosomes, and as they come off the ribosome, they fold into unique shapes, based on their amino acid sequence. Typically, they fold up into a ball-like shape with amino acids that like interacting with water on the outside (blue, green, purple) and those that hate water in the interior (yellow, orange, brown). The hydrophobic (water-hating) amino acids in the above figure are L, V, I, A, M and Y.

Importantly, while the vast majority of enzymes are proteins, not all proteins are enzymes. For example, some proteins are hormones. An example is insulin, which is made in the pancreas, travels in the blood stream (which is why it is called a hormone) and binds to insulin receptors present on the surface of all the cells in our body. This binding changes the shape of the insulin receptor (which is also a protein) and this, in turn, triggers a series of reactions within the cell to increase the number of glucose receptors on the cell's surface. These glucose receptors then bind and bring glucose into the cell for energy. So, when you eat a meal high in carbohydrates (e.g., bread, rice, pasta), you break the carbohydrates down to glucose in your small intestine and then take the glucose into your blood stream. A small number of these glucose molecules binds to receptors in the pancreas to cause the release of insulin into the blood and the insulin, in turn, increases glucose receptors on the surface of all the cells in your body to take up the glucose circulating in the blood stream for later use as an energy source.

Other proteins have structural roles. Collagen, for example, is the most abundant protein in our body and is an insoluble, three-stranded, rope-like protein that exists outside of our cells, giving strength to skin, cartilage and bone. If you have ever eaten Jello, you have eaten broken down bovine (cow) collagen, supplemented with tons of sugar, artificial flavors and dyes (since collagen is tasteless and colorless).

Now when an animal eats a plant or another animal, it breaks down the proteins in its food all the way down to single amino acids by breaking the peptide bonds that hold them together. In humans, this takes place in the stomach and small intestine. The single amino acids then travel through the blood to all our cells, where our cells take them up to use them as building blocks to make our own proteins (see below). Some proteins are difficult to break all the way down to single amino acids and if peptides (typically greater than 4 amino acids in length) get into our blood stream they can trigger food allergies if our immune system recognizes them as foreign and makes antibodies that bind to them. But I digress!

A typical human cell has about 10,000 **different** proteins in it. Most of these proteins have between 1,000 to 10,000 copies (although some proteins have as few as 10 and others as many as 500,000 copies) so that a human cell has a total of about 42 million protein molecules inside of it. These protein molecules are all being jostled about because they are constantly bombarded by the jiggling water molecules (Brownian movement) that make up most of the molecules in any cell. The higher the temperature, the faster the molecules jiggle around. The more they jiggle, the quicker that enzymes bump into their substrates. Thus, reactions happen faster at higher temperatures (within the range of just above freezing to about 40°C). Above 40°C, some proteins start to denature (lose their normal shape and so can no longer perform their function…an example is the egg white protein, ovalbumin, going from a clear liquid glob to hard egg white). Importantly, every cell has a plasma membrane surrounding it, made of fat, to keep all these mostly water-soluble protein molecules from leaking out and floating away.

OK. Time for a break so you can think about the different types of cells and the critical role that proteins play in all living things! You may want to re-read this past section since it is pretty dense.

Our Universe and How We Got Here

Part 2

This is it! You made it to the Central Dogma! This pathway is present within the cells of all living things today and confirms we are all part of one big, beautiful family.

DNA neither cares nor knows. DNA just is. And we dance to its music **Richard Dawkins**

(c) The Central Dogma

We have just covered the fact that proteins do all the heavy lifting to make life possible. This is all fine and good but where do all these proteins come from? This brings us to the awesome Central Dogma! **The "Central Dogma" is to life what fusion is to the stars!** Simply, it states that the hereditary material, DNA, gets parts of it (its genes) copied (transcribed) into messenger RNAs (mRNAs) and then the mRNAs get translated into proteins. But before we go into detail on the Central Dogma, I have to spend a few minutes describing DNA. There are several ways to depict DNA, each with specific advantages and disadvantages **(Fig 11)**.

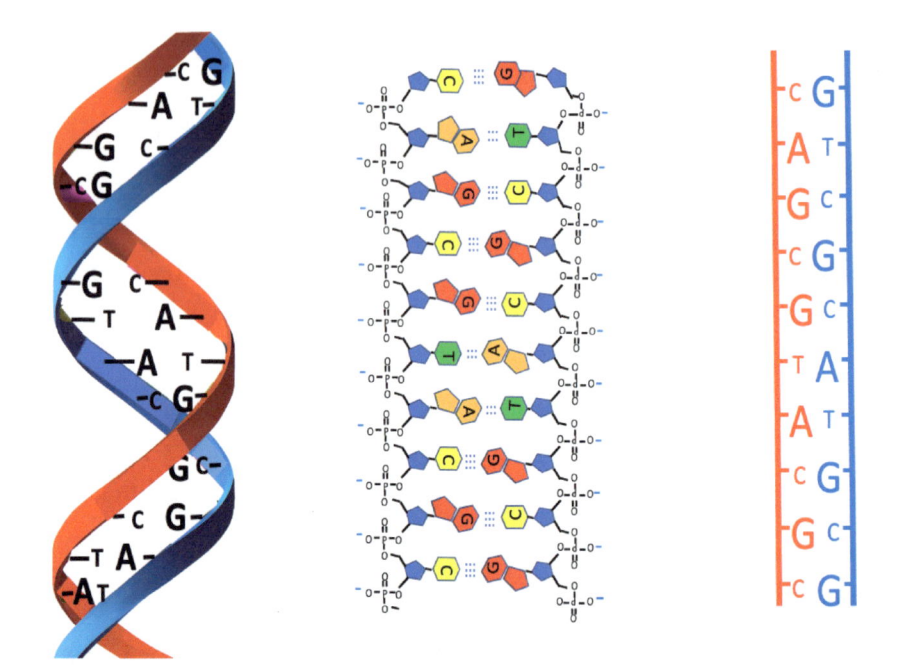

Fig 11. The structure of DNA. The left-most panel highlights the fact that DNA is a double helix with the large bases, adenine (A) and guanine (G) always interacting with the small bases thymine (T) and cytosine (C), respectively, within the hydrophobic (water hating) interior of this staircase. The middle panel, which shows the atoms in more detail, highlights the alternating sugar (deoxyribose, shown as a blue pentagon) and phosphate (PO_4) backbone that make up the red and blue bands shown in the left-most panel. Also highlighted are the hydrogen bonds holding the bases together. Because there are only 2 hydrogen bonds holding A to T and 3 holding G to C, G-C rich DNA is more stable than A-T rich DNA. The simplest way to illustrate DNA is shown in the right-most panel and highlights the fact that by having a large base always pairing with a small base, the width of the "railroad track" is always constant. Because this way of depicting DNA is the easiest to illustrate DNA replication and how mutations arise, we will be using it to illustrate these processes.

Now let's get back to the Central Dogma, where specific stretches of DNA are copied into RNA. One type of RNA, messenger RNA (mRNA), then goes to factories (ribosomes) in the cell where it is used as a blueprint to make proteins (see **Fig 12**). Now I said in the Overview that I would try and "not lose sight of the forest for the trees" but I must go into a little detail about the Central Dogma, in part because it is important in order to understand Chapter 5 where I talk about how life may have started (biogenesis) and, in part, because it gives some insight into the complexity of life and, most importantly, because it holds true for **all** living things!

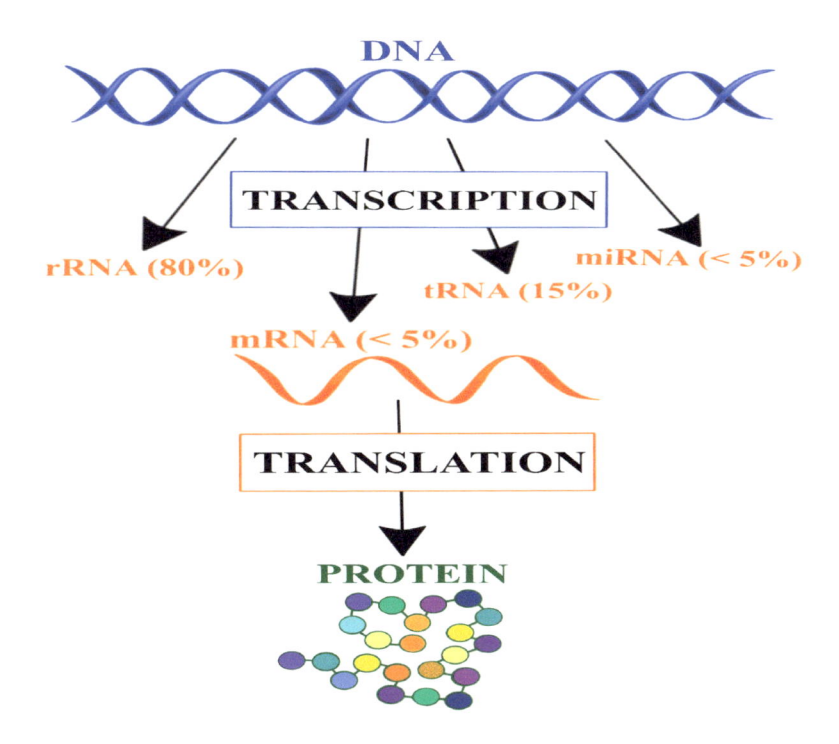

Fig 12. The Central Dogma. DNA is made up of two strands that are coiled around each other and held together by

hydrogen bonds. During transcription, parts of the DNA get copied (transcribed) into either ribosomal RNA (rRNA), messenger RNA (mRNA), transfer RNA (tRNA) or a number of regulatory RNAs, including microRNA (miRNA). Shown is the relative % of each type of RNA in a typical human cell. Only mRNA goes to the ribosomes (which are made of both ribosomal proteins and rRNA). The ribosomes then use the mRNA as a blueprint to make a protein.

OK. Let's get started! Looking back at **Fig 8**, you can see that all the different cell types, from the small prokaryotes to the large eukaryotes have DNA (typically circular in prokaryotes and linear in eukaryotes). DNA exists as an incredibly stable double stranded helix (this stability is why we can recover DNA from long dead organisms). Although DNA houses all the information to make the cell, it is important to remember that it cannot do anything by itself, i.e., it has no enzyme activity and is just a set of instructions! What makes one cell different from another, whether we are comparing a bacterial cell to a human cell, or a human liver cell to a human brain cell, etc., is the kind of proteins the cell makes. A simple analogy is that DNA is like a library. We humans have two copies of approximately 20,000 books (genes) in our library, one copy from our mother and one from our father. But to be more accurate with our analogy, let's say the books are written in a very simple 4-letter language composed of 4 bases (see **Fig 11**), adenine, thymine, guanine and cytosine, abbreviated as A, T, G and C, respectively. Now, you can't take the books out of the library, but you can make an RNA copy of a book you want (we call this transcription). The copy you make (the mRNA), contains the same

4 bases (except the T is replaced with a uracil (U), a small base very similar to T) and goes to one of many ribosomes (there can be 20,000 ribosomes in a bacterial cell and over a million in some eukaryotic cells)! At the ribosome, the mRNA gets translated into a 20-letter language with each letter in this 20-letter language being an amino acid (this translation converts mRNA into protein). Interestingly, around 1950, when it was first proposed that DNA might be the genetic material, most scientists laughed because they thought DNA with only 4 different bases was too simple to code for 20 different amino acids. And of course, this would be true if a single base specified one amino acid. However, it turns out that it is a sequence of 3 bases that specifies each amino acid. So, for example, an mRNA having the sequence AUG codes for the amino acid, methionine (single letter code = M). A sequence of 3 bases with a choice of the 4 different bases in each spot gives more than enough complexity (4x4x4 = 64 combinations) to code for 20 amino acids, and in fact there is redundancy, with more than one 3-base mRNA sequence sometimes coding for the same amino acid. Once the mRNA gets properly fitted into a ribosome it is ready to get translated into a protein. "Translated" is a good word since we are switching from a language using 4 different bases (used by DNA

and RNA) to one that uses 20 different amino acids (proteins).

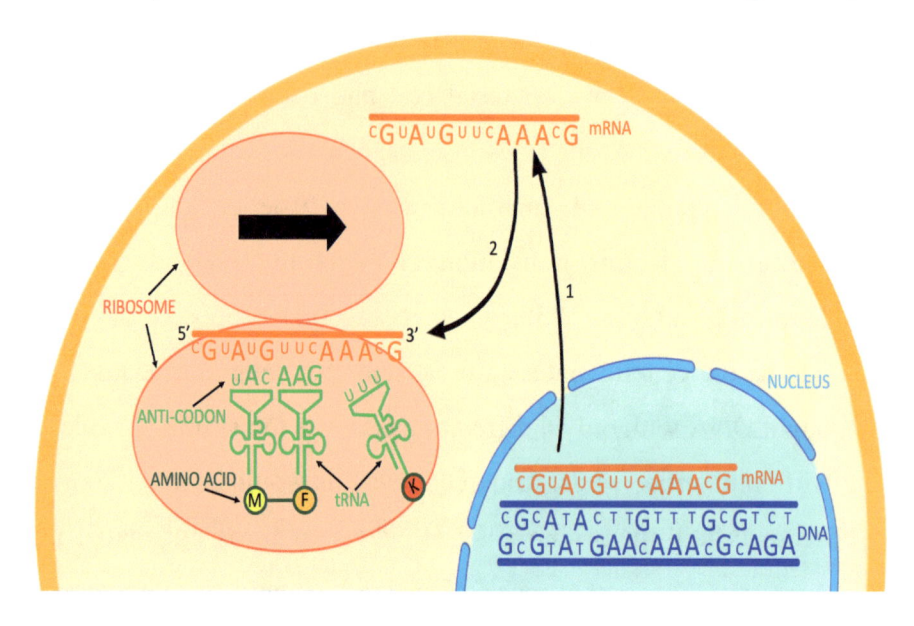

Fig 13. Details of The Central Dogma. Both DNA and RNA, in the lower right corner, are made up of 4 bases. (1) After an mRNA is made, with a nucleotide sequence that is complementary to the DNA sequence it just copied, it leaves the nucleus (in eukaryotes) and enters the cytoplasm where it (2) binds to a ribosome, made of a large and small subunit (shown in light red, on the left side of the figure). The ribosome attaches to a specific base sequence at the start of the mRNA. An average mRNA is approximately 1,000 bases long, so I am showing a **very** abbreviated version of an mRNA. Once attached, the ribosome starts moving along the mRNA in the direction indicated by the big black arrow, bringing the appropriate tRNA (shown in green) with its amino acid attached. Importantly, there is a lot of energy and a lot of enzymes (and non-enzymatic proteins) needed to make a protein. For example, there is a specific enzyme involved in attaching the right amino acid to the right tRNA, so 20 different enzymes there alone! Importantly, before a ribosome is finished moving along the mRNA to make a complete protein, another ribosome can attach itself to the front of the mRNA vacated by the first ribosome and it starts making another copy of the protein. This allows for a rapid increase in the number of copies of a specific

protein! The single letter codes for M (yellow), F (orange) and K (red) stand for methionine, phenylalanine and lysine, respectively.

For those of you who just read the Fig 13 figure legend, I apologize! This was likely too much detail and new material for those who have not been exposed to biology before. But unglaze those eyes! The most important part is about to come, **transfer RNA** (tRNA). This really is the most important player in the Central Dogma because it is the bridge molecule that translates the 4 base DNA/RNA language into the 20 amino acid protein language. You can think of tRNA as a biological "Rosetta stone" since one end of this tRNA molecule has a 3 base sequence that is complementary to the codon on the mRNA (and so is called the anti-codon). The other end of the tRNA has the corresponding amino acid attached (see **Fig 13**). There are over 20 different tRNAs and each binds only one of the 20 specific amino acids. To ensure that the right amino acid gets attached to the right tRNA there are over 20 specific enzymes called aminoacyl tRNA synthetases. Each aminoacyl tRNA synthetase picks up a specific amino acid and attaches it to the right tRNA (by recognizing a specific base sequence in the tRNA, close to where it will attach its amino acid). Once two tRNAs are in position within the ribosome, a peptide bond is formed between their two attached amino acids to join the two amino acids together (see **Fig 13**). Interestingly, the formation of this peptide bond is catalyzed by **an RNA** within the ribosome (ribosomes are made up of approximately 60% RNA and 40% protein) and not by a protein enzyme!! So, this is one of the rare instances in which an RNA molecule has enzymatic activity.

(d) Genetics and Epigenetics

When a cell divides to form two cells, the DNA in the parental cell has to be copied exactly so that both daughter cells have the same sequence of DNA bases. To make an exact copy, the zipper holding the two strands of DNA together (the double helix) opens up by breaking the connecting hydrogen bonds. Complementary bases are then brought in to make exact copies (**Fig 14**). Many cells in our body are dividing. For example, in our bone marrow we humans make 2 million red blood cells every **second** (so don't let anyone ever call you lazy again!!!). Of course, this synthesis of new red blood cells replaces the 2 million red blood cells that are dying every second, so the number of mature red blood cells stays constant.

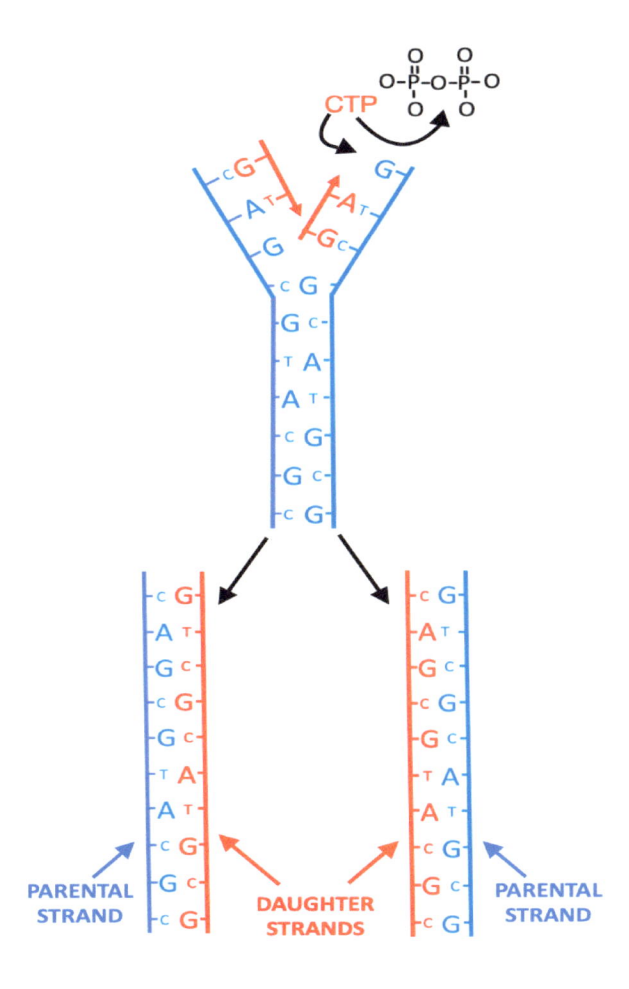

Fig 14. DNA replication. Before a cell divides to form two daughter cells, the DNA must be copied exactly so each daughter cell has a complete set of instructions. To do this, the two strands of DNA unzip (i.e., the hydrogen bonds holding the complementary bases are broken) and complementary bases (shown in red) are put in, as shown in the top panel. The backbone of DNA is actually composed of a negatively charged phosphate group alternating with a simple sugar (a 5-carbon sugar called deoxyribose) as shown in the middle panel of Fig 11. The nucleotides (ATP, GTP, CTP and TTP) coming in to form the new daughter strand have, as part of their structure, a deoxyglucose and three phosphate groups. The two terminal, high energy phosphate groups on the newly arriving nucleotides are

broken off, as shown for CTP in the figure, to provide the energy needed to synthesize the new DNA strands. Once an exact copy of the DNA is generated, the cell can divide, and each daughter cell gets exactly the same instructions (DNA) as the original cell. Notice that, as mentioned earlier, a large purine base (A or G) always pairs with a small pyrimidine base (C or T), keeping the width of the double-stranded DNA constant.

Genetics is the study of genes and how genetic variation is passed on from one generation to the next. So, genetics concentrates on elucidating (figuring out) the sequence of the bases in DNA. Talking specifically about humans, we have about 6 billion bases in our DNA, which is scattered over 46 chromosomes (23 from our mother and 23 from our father). Interestingly, if we just put these 6 billion bases end to end, the double-stranded DNA would stretch over 6 ft (2 meters) in length. Although this double stranded DNA helix would be too thin to see with the naked eye, it is still amazingly long relative to the cell nucleus it has to get stuffed into. All DNA, regardless of the organism, would actually prefer to be all stretched out like a very thin pole, as just described, because it is covered with very negatively charged phosphate groups (see middle panel, **Fig 11**) that repel each other. What allows eukaryotic DNA to twist up into a very compact structure that can fit into the nucleus of a eukaryotic cell is a covering of positively charged proteins, called histones (which are attracted to and neutralize the negative phosphate groups of the DNA). Archaea, like eukaryotes, also use histones. Bacteria, however, don't use histones but much smaller positively charged molecules called polyamines. This difference is one reason it is thought that we, and other eukaryotes, evolved from Archaea rather than bacteria. When eukaryotic DNA is tightly covered with histones it can become so tightly condensed that the enzyme responsible for transcribing DNA into mRNA (i.e., RNA polymerase) can't get in to make an mRNA (and thus a protein).

When proteins called transcription factors come into the nucleus to tell a gene to "turn on" and make an mRNA for subsequent protein synthesis, they have to drag enzymes with them that make the positively charged histones less positive. This reduced positive charge lowers their attraction for the negatively charged phosphate on the DNA and thus opens up the DNA, so it is more accessible to the transcription factors, and allows RNA polymerase to make an mRNA. Such modifications of the histones are called epigenetic ("in addition to genetics") changes, since they do not involve changes to the DNA **sequence** itself). Another example of an epigenetic change is the methylation of certain bases (the addition of a methyl (CH_3) group, mostly onto the C bases) in the DNA itself. This also affects the ability of genes to be transcribed (but doesn't change the DNA sequence) (see **Fig 15**).

These epigenetic changes can be passed on to daughter cells after cell division and can even affect the progeny (next generation) of living organisms (i.e., they are inheritable). There are many fascinating examples of this. For example, it has been shown that the reason certain smells make us feel uneasy could be due to epigenetics. A study found that, "Mice whose father or grandfather learned to associate the smell of cherry blossoms with an electric shock became jumpier in the presence of the same odor and responded to lower concentrations of it than normal mice" (23). So perhaps, irrational fears are a result of epigenetics

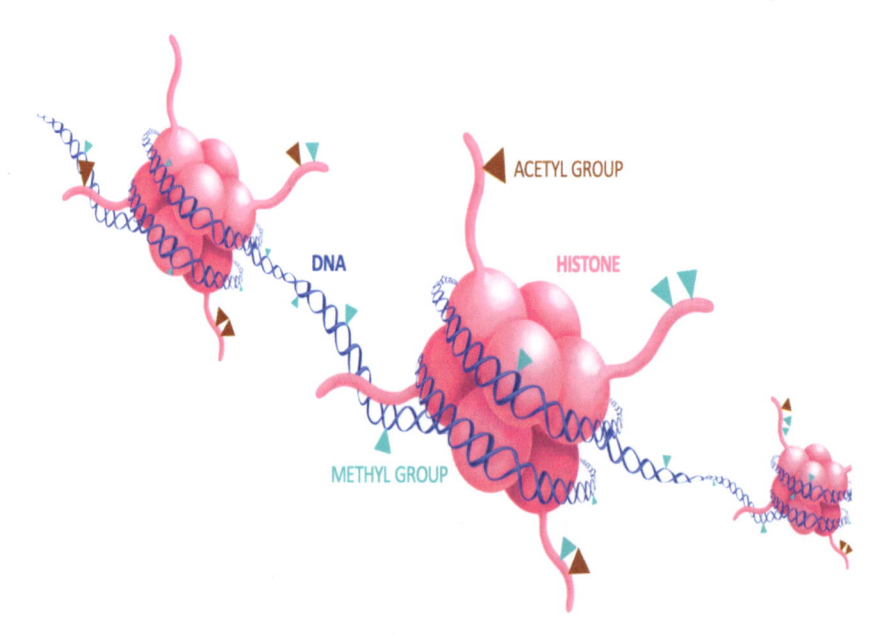

Fig 15. Epigenetic changes. Both the DNA itself and the histones (shown as large red balls) that are encircled by DNA can be modified. Specifically, some Cs and some As in DNA can get a methyl group (CH_3) attached (shown as a green triangle) and histones coating the DNA can also get a methyl (green triangle) or acetyl (CH_3CO) (brown triangle) group attached. These changes to the positively charged histones alter their attraction to the negatively charged DNA, exposing (opening up) the DNA for transcription or compacting it (closing it down) to prevent transcription.

As a last note on epigenetic changes, DNA damaging agents have long been known to cause aging, and it was always thought that this was because they increased DNA mutations and it was these wrong bases that led to aging. But a 2023 paper has revealed that the aging may be due, at least in part, to changes in the epigenetic landscape during repair (24).

(e) Post-translational Modifications

Let's pretend for a moment you are a cell in your blood stream. But before we go there, I have to mention that your blood is red because most of the cells in your blood are red blood cells, and these cells are red because they contain the red colored, oxygen transporting protein, hemoglobin. Although far fewer in number, you also have non-hemoglobin containing cells, appropriately called white blood cells. These cells are there to protect you from invaders (are part of your immune system). One type of white blood cell is called a monocyte and its job is to look for any damage to the inside of blood vessel walls and to alert other immune cells in the blood if a bacterial, viral of fungal invader has come into the blood stream. OK! So, let's go back to pretending you are a cell in your blood stream, and specifically a monocyte. You are being pushed along in your blood stream by the pounding of your heart, and you bump into some bacteria that entered your blood through a cut. If you don't respond quickly to these invaders, they will multiply and kill you (bacteria can double their number every 20 to 30 minutes if there are enough nutrients present). Being a monocyte, you have receptors (proteins) on your cell surface that recognize the invader and bind to it. But how do you react quickly? This is where post-translational modifications come in (i.e., after mRNAs have been translated into proteins). These are rapid changes to proteins that are already present within the cell. One example of a post-translational modification is putting phosphate groups on specific amino acids within an existing protein to increase/decrease enzyme activity or affinity for other molecules. This small change to a protein occurs within seconds of a cell surface receptor on the monocyte binding a molecule (in this case on a bacterial cell wall). This binding triggers cascades of protein modifications within the monocyte. A simple analogy is a group of fire fighters passing buckets of water to put out a fire, with the bucket of water being a phosphate group in this example. Think of lines of fire fighters going off in different directions within the monocyte. Some lead to the activation of transcription factors that go into the nucleus to turn on the genes for inflammatory proteins that are going to be secreted from the monocyte to alert and recruit other members of the immune system. However, you should keep in mind that once a transcription factor binds to genes to "turn them on", it takes 1 to 3 hours to make significant levels of inflammatory proteins. So, to react quickly, you also activate post-translational cascades (lines of fire fighters) to trigger engulfment (phagocytosis) and digestion of the bacterium within the monocyte. So remember, post-translational modifications allow for very fast changes in the behavior of a cell because they occur on proteins **already present** in the cell (**Fig 16**); you don't need protein synthesis! Interestingly, it has recently been shown that even Mycobacteria, one of the simplest forms of bacteria, use post-translational modifications, suggesting that this mechanism may have been developed very early on in evolution to allow for quick responses to environmental changes (25).

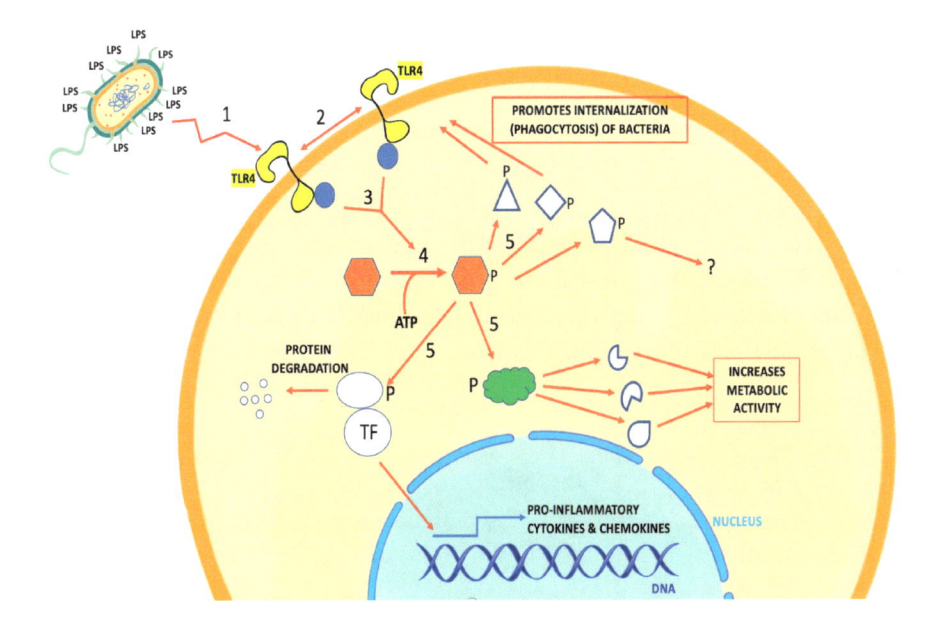

Fig 16. Post-translational Modifications. These are small changes to proteins that are already present in a cell that allow the cell to respond quickly to environmental changes. In the example shown, LPS (lipopolysaccharide) on the surface of Gram-negative bacteria (e.g., E. coli) is bound/grabbed by a cell surface protein (a Toll-like receptor called TLR4 in this case) on a human monocyte (a white blood cell) in the blood stream (step 1) and this triggers the dimerization of the TLR4 on the surface of the monocyte (step 2). This coming together of two TLR4 receptors activates an associated protein kinase, shown as a blue circle (step 3), and this enzyme initiates many signaling cascades inside the monocyte, by transferring a phosphate group (PO_4) (which is negatively charged) from ATP (discussed in the next chapter) to specific proteins in the

cell (steps 4 & 5). Some of these proteins are also protein kinases (all protein kinases in this example are colored). These phosphorylations change the shapes, possible interactions and/or enzymatic activity of proteins, leading to (a) changes in the metabolic activity of the monocyte, (b) the engulfment of the bacterium and (c) the activation of transcription factors (specific proteins that go into the nucleus to "turn on" and "off" genes). In the example shown, the transcription factor (TF) is held by a protein in the cytoplasm, making it inactive. Phosphorylation of the TF-holding protein leads to its degradation, resulting in the release of the TF. Once released, the TF goes into the nucleus. Transcription factors bind to DNA and stimulate the transcription, in this case, of pro-inflammatory chemokines (i.e., proteins which recruit other immune cells) or pro-inflammatory cytokines (i.e., proteins which activate other immune cells to destroy the invading bacteria). Shortly after these phosphorylations occur and trigger the biological changes shown, other enzymes called protein phosphatases remove the phosphates to bring the cell back to a resting state.

If you feel that this section was really dense, you are right, and I highly recommend you read it once or twice more to fully understand it. If you feel, on the other hand, that you are really dense, I recommend you just skip to the recap below to get the salient (important) points (Ha! Ha!!).

To recap, all life today is composed of cells, with most living things being made up of just one cell (unicellular). The key feature of living organisms is their enzymes, which help convert molecules (substrates) in their environment into molecules (products) they can use to survive, grow and generate offspring. The blueprint for these enzymes (and all proteins) is provided by the sequence of bases within DNA. All living things generate their proteins via a pathway called the Central Dogma. This pathway can be summarized as the conversion of DNA to messenger RNA (mRNA), to protein. Important facts to remember are that (a) DNA by itself is just a "book of instructions" with no enzyme activity, and so it can't generate a cell by itself, (b) proteins can't be translated back into RNA (it is a one-way trip from mRNA to protein) and (c) tRNA is the critical "translator" molecule of the Central Dogma! For a cell to change its behavior **quickly** in response to a change in its environment, it does not rely on new protein synthesis, which takes a few hours, but does so by modifying proteins that are already present in the cell (post-translational modifications). So, when you encounter a bear in the woods, both you and the bear can respond almost instantly because of post-translational modifications.

EVOLUTION

Chapter 4

Part 1

Now that you have some basic biology under your belt you are ready to begin the long journey from single-celled prokaryotes to us. In this first section you are going to learn that all living things need energy to survive, how they generate it and how they store it for later use.

There are two types of people in the world, those who want to know and those who want to believe   **Friedrich Nietzsche**

First of all, I would like to congratulate you for making it through the last chapter, especially those of you who haven't been exposed to very much biology (Playboy magazine doesn't count!). I know the Central Dogma was a heavy slog. And for those with a thorough knowledge of biology, I apologize for oversimplifying. But I have tried to limit myself to information that you will need to understand this chapter on evolution and the next on biogenesis.

Secondly, as you likely know, the concept of evolution challenges the Judeo-Christian belief in creation. For those who take the Bible literally there is a general belief that God created the universe 6,000 to 12,000 years ago. In fact, in the 1600s an Irish Anglican archbishop, James Ussher, calculated that October 23rd, 4,004 B.C. was the time the universe was created by counting all the "begats" in the Old Testament. This, of course, suggests that every living thing, our Earth and all the stars came into being 4,004 years ago. As well, in Europe from the 1600s to the 1800s it was generally accepted that God created all living things in a perfect, immutable (non-evolving) fashion. The standard counter argument by Christian clergy to any support of evolution, with species appearing and disappearing, is/was "Why would God create a species just to destroy it?" This generally held belief was severely shaken when dinosaur bones were first discovered in the early 1800s. If the world was only 4,000 - 6,000 years old, this meant that dinosaurs and people co-existed. Can you imagine a cave man **not** painting pictures of dinosaurs on the inside of their caves if they had co-existed? Nonetheless, these theological concepts influenced, frightened and severely restrained scientists in Europe during this time. It must be remembered that many of these scientists were brought up to be deeply religious themselves and were studying nature to "understand and get closer to God". It is also important to keep in mind that we are all products of our time (with the possible exception of Leonardo da Vinci), whether we like it or not. Even Charles Darwin, the great English naturalist who proposed in the 1800s that all life evolved over time from a common ancestor by natural selection, was terrified of publishing his "On the Origin of Species". For most of his adult life he suffered terrible, long bouts of severe illness, including heart palpitations and severe abdominal problems. It is likely that his discoveries and the anguish they caused him and his religious wife, Emma, contributed to his poor health (26).

*B*ased on our current understanding of evolution it is thought that life arose on Earth from a simple single-celled ancestor around 4 billion years ago. Multiple lines of evidence point to this as the logical explanation for the diversity of life that exists today, including fossil records, comparative anatomy, embryonic development and, most importantly, as I touched on in the previous chapter on the Central Dogma, a common biochemistry that has diverged over time. As Theodosius Dobzhansky said in 1973, "nothing in biology makes sense except in the light of evolution." For example, if I were designing human beings from scratch, they would not have (a) residual, non-functional tails (coccyges), (b) their 46 chromosomes strewn with the DNA of viruses that infected us during evolution so that about 40% of our total DNA now consists of residual bits of viral DNA or (c) some of the 20,000 genes we possess mutated into inactivity, so that they no longer code for functional proteins. Related to this last point, unlike most bacteria, humans can no longer make functional enzymes to catalyze the synthesis of all 20 amino acids needed to build proteins and so we must get 9 amino acids from our diet (i.e., these are called "essential" amino acids). However, as long as we can get these 9 amino acids from our diet, this is not a bad thing since it saves us having to expend energy (ATP) synthesizing the enzymes needed to make these amino acids. I could go on about the ridiculous design of the human eye with its silly blind spot, but all these things make sense if you realize that life began simply and there are only so many ways to get here from there.

It is also the height of immodesty to think we are somehow distinct from or superior to all other living things rather than part of a wonderfully diverse family of different species. At this point, I should define the term "species". For higher eukaryotes, a species is a group in which all members can mate to produce fertile offspring. For prokaryotes, which reproduce asexually, a species is defined as a group having a very similar DNA sequence. Accepting that all species today originated from the same single-celled ancestor should instill a little more empathy for our fellow life forms and encourage us to treat them better in the future. Related to this, all living things today, including all humans, are survivors of 4 billion years of evolution. So we should all give each other, our pets, the birds in the trees, the bugs in the grass and ourselves a pat on the back. Said another way, you, dear reader, are the unlikely product of 4 billion years of successful reproduction, like a long unbroken but very tenuous thread. This makes your very existence all the more remarkable and worthy of celebration. The reason I say this is that the lives of individual species are very precarious and often transient, with over 99% of all the species that have ever lived no longer being around. On the other hand, life itself is **incredibly** resilient and ubiquitous on Earth. Wherever there is water, even if it is two miles underground in tiny fissures within solid rock and in total darkness, there is life (27).

In the paragraph above I was thinking of saying life is incredibly "tenacious" but I decided not to use this word since it suggests that living things inherently have a strong "desire" to survive and this takes us into hazardous, metaphysical waters (i.e., the philosophical pursuit of the nature of reality and causality). However, if you put a tiny sample of pond water under a simple light microscope, you see a tremendous array of single-celled and multicellular life forms, many swimming via tails (flagella) or many little hairs (cilia), eating each other or nibbling on chlorophyll-containing plant-like material. Staring at this for a while, you get the sense that even single-celled creatures with no known pain receptors (no nervous system) seem to actively avoid being eaten. On the other hand, one can argue that 4 billion years of evolution have simply selected for single-celled organisms that activate intracellular pathways that promote avoidance movement when nibbled on. To remain somewhat philosophical for just one more sentence, the desire to stay alive, if real, would give a tremendous selective advantage to any species, be it single-celled or multicellular.

OK. I have to say one more thing on this subject before letting it go. It has long been thought that "learning" (defined as an adaptive change in an organism's behavior as a result of experience) requires a nervous system so that connections (synapses) between nerve cells can strengthen on learning a new behavior. However, in the 1950s an all but forgotten, very meticulous scientist by the name of Beatrice Gelber showed that Paramecia, which are unicellular eukaryotes (protists) covered in cilia to help them move, "learned" to move towards a wire dipped in bacteria in order to feed on the bacteria. Importantly, she found that subsequent dipping of a wire without bacteria still attracted them (a Pavlovian response), even though these primitive single-celled organisms obviously lacked a nervous system (28). This incredible finding was largely ignored, in part because science was not ready for the idea that single-celled organisms could learn (timing is everything in life, including science) and, unfortunately, because science was still largely a "boys club" back then (and still is, in some disciplines). So, what I am trying to say is that we still have a lot to learn, even about the simplest single-celled organisms.

OK. I know I promised to let this go but now I'm on a roll. In the early 1960s when I was an undergraduate student at McGill University (yes, I'm really old) I heard a talk by Cleve Backster, an interrogator at the CIA in the USA. He claimed, by using a polygraph (lie detector) attached to plant leaves, that plants responded not only to personal damage but to the death of other living things as well. He went so far as to say even the threat/thought of harming these plants caused a response (suggesting telepathy…perhaps a bit of a stretch). He suggested that this "sensitivity" that plants displayed, which he dubbed "primary perception", was present at a cellular level in all plant and animal life (29). Although his claims were rejected at the time by the scientific community, a paper published in 2023 in the very prestigious journal, Cell, revealed that tomato and tobacco plants emit ultrasonic sounds (which we can't hear) when cut or not watered, perhaps via the popping of air bubbles in their xylem (the vasculature in plants that transports water from their roots to the rest of the plant). Interestingly, the plants made different sounds when they were not watered versus cut (30). Related to this, it has been reported that plants can respond to just the sound of a caterpillar chewing on its leaves, rather than the actual caterpillar chewing its leaves, by secreting toxins that discourage the caterpillar from continuing to nibble. Other, non-threatening sounds did not have this effect (31). Taking these findings a little further, there are anecdotal reports of personality changes following heart transplants (32). One interpretation of these reports is that the transplanted hearts "remember" their previous host.

However, before totally entering the "Twilight Zone" it is important to remember that **responsiveness is not the same as awareness** and there are more mundane explanations that must be considered. For example, it is possible that a cut or dehydrated plant has its flow of xylem disrupted by drought and stem cutting and this could result in more air bubble formation and, thus, more ultrasonic noise. So, while it may serve as a distress signal to insects and other animals close enough and able to hear ultrasonic vibrations, the damaged plant itself may not be "feeling" any distress. For those interested in whether plants can learn, please check out (33). As far as the anecdotal stories around heart transplants, all of our organs contain macrophages, which are immune cells that secrete pro-inflammatory proteins (cytokines) to fight infections as well as anti-inflammatory cytokines to promote healing after tissue damage (e.g., once the infection has been cleared). Interestingly, it has been shown that these cytokines can influence our emotional state (34) and this may account for the change in personality reported after a heart transplant. Let's just say that, as of 2023, the jury is still out on whether single cells or plants have something akin to "awareness".

(a) Life Needs Energy

OK. Let's start with what may seem like an odd and/or obvious statement. All living things need an external source of energy to survive, grow and reproduce. Philosophically speaking, what do you imagine life might be like if this were not the case? For example, what if we were more like a star, generating our own energy through something like fusion? Or what if we were born with enough energy stores to take us through our entire life rather than just until birth/germination and shortly thereafter. But all living things on Earth do need an external (exogenous) source of energy and this is what **drives evolution**, because we all have to compete for limited amounts of energy! Another way of saying this is a species survives if it gets enough energy to produce more (or equal numbers of) offspring than dies. You can't get simpler than that! All living things on Earth get the energy they need either from simple chemicals (chemotrophic bacteria and some archaea), directly from the sun (plants), or by eating other organisms (everyone else). So how do we get and use this energy from these sources?

(b) How Living Things Store and Generate Energy

When an organism gets energy from the various sources

described in the last sentence, it can store it for later within covalent bonds of various molecules. Plants, for example, mainly store the energy they obtain from the sun (through photosynthesis) within the sugar polymer, starch. Animals, on the other hand, mainly store the energy they get (from eating other living organisms) within fat. Adenosine triphosphate (ATP, that is also used to make RNA) can be thought of as the "money" that all living things make and "spend" to help them survive. They can either make this money directly when food molecules are broken down, or when they break the bonds of their storage depots (starch or fat). All living organisms "spend" their ATP by breaking off one or both of their two terminal phosphates to become energy-depleted AMP (see **Fig 17**). How this ATP is spent for survival depends on the life form, but all living things use a portion of the ATP they generate just for routine maintenance (just staying alive). On top of this energy-requiring maintenance, some life forms use their remaining ATP for chasing prey or running from predators (using ATP to power muscles) and others use it to grow (cell division). Getting big in a hurry is definitely a huge survival advantage (pun intended), since you are less likely to be eaten, but it takes a lot of ATP to both grow and to maintain a large body. So, when times are tough (i.e., when there is very little energy (food) coming in to make ATP) it is better to be small since it takes less ATP to maintain a small body. When the food supply gets really low, living things tend to minimize non-essential ATP-draining events and some can even go into a dormant-like state (e.g., hibernation) in which less ATP is needed for maintenance. Some microorganisms go all the way into a spore-like state, requiring no ATP at all, and remain viable until conditions improve.

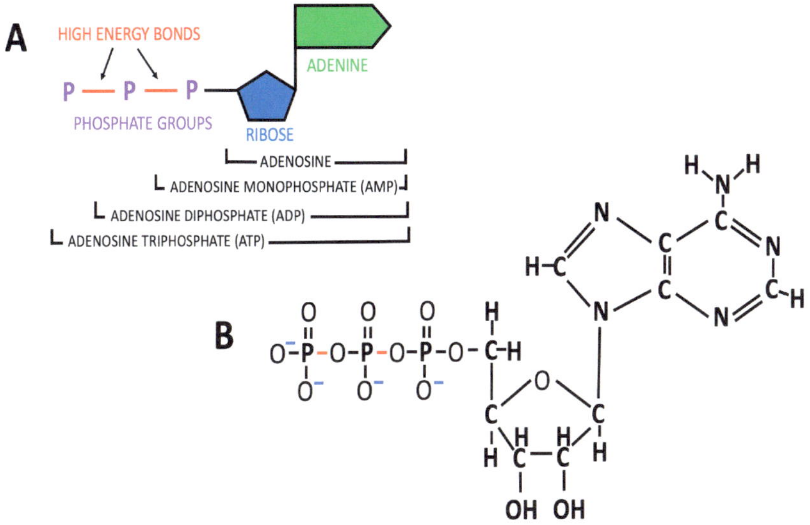

Fig 17. The structure of ATP. (A) A simple depiction of ATP, made up of the nucleic acid base, adenine (A), shown in green, the simple 5 carbon sugar, ribose, shown in yellow, and three negatively charged phosphate groups. **(B) A more realistic depiction of ATP**, showing all the atoms involved. ATP stores its energy in its outer two phosphate bonds (shown in red). The oxygen atoms in the phosphate groups are negatively charged because they pull electrons from the phosphorus atoms towards them (oxygen atoms are electrophilic, meaning they love electrons), causing an increase in the electron/proton ratio in the oxygen atoms. When the high energy phosphate bonds are broken, energy is released to allow certain enzyme-catalyzed reactions (like building macromolecules) to occur, and this converts ATP into either the less energetic adenosine diphosphate (ADP) or the totally energy-depleted adenosine monophosphate (AMP).

But how do living things make ATP? As mentioned earlier, some living things can make ATP (generate energy) from eating other living things or from sunlight. Then there are chemotrophs, which are a group of bacteria and archaea that can make ATP from simple chemicals like hydrogen, hydrogen sulfide, ferrous iron, methane or ammonia (35). These chemotrophs, especially the chemoautotrophs (which can live and grow in the absence of any organic molecules), are good candidates for being the earliest forms of life.

(c) Anabolism and Catabolism

OK. Before I bore you with how living things make ATP (I am a biochemist after all) I want to discuss the biosynthesis (anabolism) and breakdown (catabolism) of macromolecules (big molecules). Specifically, I want to talk about the synthesis of nucleic acids, carbohydrates, proteins, and lipids that all living things possess. Making these macromolecules requires energy. Catabolism, on the other hand, refers to the breakdown of these large molecules into smaller subunits, usually releasing energy that the organism can use to make ATP.

So, to start, all living organisms contain nucleic acids (DNA and RNA, which contain the **instructions** for building and maintaining an organism), proteins (which carry out the vast majority of the enzymatic and structural functions of a living organism), lipids (also called fats, which are needed to form membranes to keep cells and intracellular compartments intact and are also used to store energy) and carbohydrates (to store energy (e.g., starch) and modify proteins and lipids) (**Fig 18**). These 4 macromolecules are composed of small building blocks (subunits) that are joined together (via strong covalent bonds) to form them. For example, all nucleic acids are made up of just 5 different bases, all proteins are made up of 20 different amino acids, and all carbohydrates are made of simple sugars (like glucose). Starch, for example, which is the major storage form of energy in plants, is a carbohydrate composed of hundreds of glucoses attached end to end. Fats, on the other hand, are typically made of 3 long-chain fatty acids held together by a simple 3 carbon molecule called glycerol (see **Fig 18**).

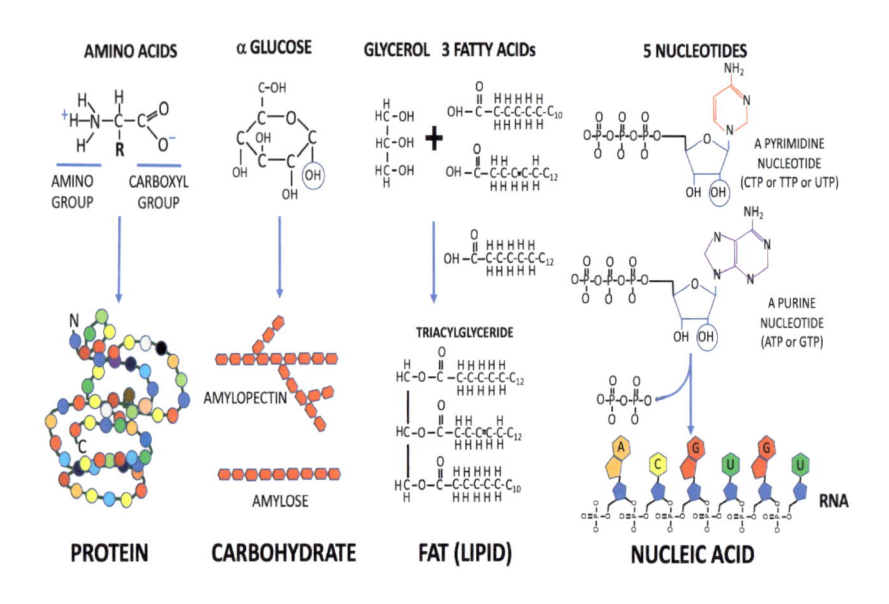

Fig 18. The 4 major macromolecules. Shown in the bottom row are the molecular structures of a typical protein, carbohydrate, fat (lipid) and nucleic acid and, in the top row, the building blocks used to make them. **Proteins**: All 20 amino acid building blocks have the structure shown in the top left panel. The only difference is the composition of the "R" group, which can be as simple as a hydrogen atom (i.e., in glycine) or as complex as a ring structure (e.g., in tyrosine). All amino acids have an amino and a carboxyl end. Since proteins are simple long linear strings of amino acids (shown as differently colored beads in the bottom left panel), the first amino acid has an exposed amino group (an "N" in the bottom left panel) and the last amino acid, an exposed carboxyl ("C") group. Proteins typically fold up into ball-like structures so that amino acids with R groups that like water (hydrophilic) are on the outside interacting with the watery medium, and those with R groups that hate water (hydrophobic) are hidden inside the ball of protein. **Carbohydrates**: The carbohydrate building block is usually glucose. Alpha (α) glucose is shown (with the circled hydroxyl group (OH) pointing down). This is the building block of starch. Beta (β) glucose has this OH pointing up and is the building block of cellulose (which comprises the cell walls of plants and is the most abundant organic molecule on Earth). Starch can be composed of either branched

chains of glucose (amylopectin and glycogen), which is easy to break back down to glucose, or amylose, which is unbranched and thus stacks together to form water-insoluble aggregates that are resistant to digestion (hard to break down). **Fat**: Triglycerides are the most abundant type of fat in humans (and other vertebrates) and are composed of glycerol (a simple 3-carbon (C) molecule, attached to 3 fatty acids. The fatty acids are just long linear carbon chains, usually ranging in length from 16 to 22 carbons. If 2 neighboring hydrogens are removed, a double bond is formed (shown in the 2nd fatty acid above). Fatty acids with these double bonds are called unsaturated (i.e., not saturated with hydrogens). If the hydrogens are lost from the same side of the C chain, the remaining hydrogens are side by side (in cis, as shown in the figure) but if the hydrogens are removed from opposite sides, the remaining hydrogens are in trans and these trans fatty acids increase coronary heart disease. Because all long chain fatty acids are just composed of carbon and hydrogen, these triglycerides hate water and are stored as fat droplets in fat cells (adipocytes). **Nucleic acids**: Nucleoside triphosphates (nucleotides) are the building blocks of both RNA and DNA. Purine nucleotides (ATP and GTP) are larger than pyrimidine nucleotides (CTP, UTP and CTP) and, in DNA, a large base always hydrogen bonds to a small base (i.e., A=T and G≡C). Shown in the figure are the building blocks for RNA, which is single-stranded, and has the circled hydroxyl (OH) group within the ribose moiety (shown in blue). For DNA, which is double-stranded, these building blocks have this OH group removed (the ribose group is converted to a deoxyribose group). Shown in the bottom panel is the structure of RNA. Of note, in both RNA and DNA, the bases are held in position by alternating phosphate and sugar (ribose or deoxyribose) residues. Importantly, the removal (hydrolysis) of the two terminal phosphate groups from the building blocks provides the energy required to construct the nucleic acids.

Plants have far lower levels of fat and far higher levels of carbohydrate than animals. These higher levels of carbohydrate in plants are, to some extent, because of their cell walls, which are made of cellulose, i.e., many beta (β)-glucoses attached end-to-end by bonds that can only be broken by certain bacteria. However, the main reason for the far higher carbohydrate level in plants is because their main storage form of energy is starch, i.e., alpha (α) glucoses attached end-to-end by bonds that are easily broken by both plants and animals. Animals, however, have chosen fat over starch as their main storage form of energy. So why do animals primarily store their energy as fat while plants store theirs as carbohydrate? Well, one reason is that when fat is broken down it generates twice as much energy (ATP) per gram than when you break down carbohydrates (or proteins). So, fat gives you the same amount of energy as carbohydrate at half the weight. Put another way, would you prefer a 25-pound or a 50-pound backpack when you are being chased by a bear? The reality is even worse than this because starch, (as well as simple sugars) needs to have a "shell" of water around it to keep it in solution. Thus, your backpack would be even heavier if you were mainly using carbohydrate as your energy source because of the water it binds up. This extra weight doesn't matter as much to a plant since it never has to run away from a bear.

As an aside, you are all familiar with the 3 major food groups; proteins, carbohydrates and fats, but never hear about the 4th major macromolecule, nucleic acids. That is because it only constitutes about 1% of the macromolecules in plants and animals and so is not considered a major food group.

(d) Central Metabolism

How ATP is actually generated from the breakdown of food (organic compounds) brings us to what is called central metabolism, which is shown in **Figs 19A & B**. Please look at **Fig 19A** to follow what I am now about to say because it gets a little complicated. Don't let all the detailed circuitry in **Fig 19A** throw you. The details are not important, but you should know, however, that I am showing you a very simplified version of central metabolism. Each step indicated by an arrow requires a specific enzyme to convert one molecule into the next and I have not included the names of any of the enzymes involved. For example, once glucose is brought into a cell via a glucose transporter (a protein embedded in the plasma membrane) it is rapidly phosphorylated to glucose-6-phosphate (G-6-P) by an enzyme called hexokinase. The newly attached phosphate makes glucose negatively charged (phosphate groups are very negatively charged) so it can't easily diffuse through the lipid cell membrane to get out of the cell. Central metabolism starts with glycolysis. Glycolysis is a 10-step process in which a six-carbon (6-C)-containing glucose is converted into two molecules of the 3-C-containing pyruvate. Each step requires a different enzyme, with the entire process taking place in the cell's cytoplasm. This process generates two ATPs per glucose molecule. The newly generated pyruvate then has an important choice to make, based on oxygen levels. If there is very little oxygen (hypoxia), it gets converted into lactate and is secreted from the cell.

If, on the other hand, there is sufficient O_2, pyruvate goes into one of the many mitochondria present in eukaryotic cells and enters what is called the tricarboxylic acid (TCA) cycle (also known as the citric acid or Krebs cycle). Once there, it loses a carbon, as carbon dioxide (CO_2), and becomes a 2-C acetyl group, attached to coenzyme A (CoA) and so is called acetyl-CoA. The job of CoA is to carry (transport) this acetyl group to the right place to get attached to a 4-carbon molecule called oxaloacetate. This results in the generation of the 6-C citrate molecule (look at **Fig 19A**). In the TCA cycle this 6-C citrate loses the two carbons it got from the acetyl group, as CO_2, within a 6 step, enzymatically catalysed process, breaking it back down to oxaloacetate so that it can accept another acetyl-CoA group to go around again. So, what is the purpose of this little cycle? If you look closely at **Fig 19A**, you will see that for every pyruvate that enters the mitochondria, we generate 3 molecules of CO_2, that we breathe out, and 1 ATP. More importantly, we generate 4 hydrogen atoms with very high-energy electrons. These high-energy electrons are carried from the TCA cycle to the electron transport chain by two critical transporters (carriers) called, nicotinamide adenine dinucleotide (**NAD**) and flavin adenine dinucleotide (**FAD**) (shown in **Fig 19B**).

Specifically, after each TCA cycle, the 2 **NADHs** and 1 **FADH$_2$** that are generated in the mitochondria stay in this organelle but travel to one of its many inner membranes. Once there, the high-energy electrons leave their carriers and activate a series of protein pumps (shown in blue) embedded in the inner membrane. These pumps push hydrogen nuclei (which, if you remember from Fig 1, are just positively charged protons) into the space between the inner and outer membrane (leaflet) of the mitochondria (**see Fig 19B**). As the high-energy electrons get passed from one pump to the next, they lose a little bit of energy in activating the pumps to push out protons. The pile up of protons in the space between the inner and outer leaflet of the mitochondria (or between the plasma membrane and the cell wall in prokaryotes) generates an unequal distribution of these protons (which is a form of potential energy) so that the protons "want" to come back in so they can become equally distributed again. Their coming back in (via a "revolving door-like" enzyme called ATP synthase) provides the energy to convert a lot of ADP into ATP (36 ATPs for every glucose that is taken into the cell, highlighted in yellow in **Fig 19B**). So, if there is oxygen present in a cell, we can generate a lot more energy using the TCA cycle and the electron transport chain than using glycolysis alone (which only generates 2 ATPs/glucose). So why do we need oxygen? When the high-energy electrons excite the proton pumps to push protons out of the mitochondria, they eventually become energy-depleted and combine with both the protons that are coming back into the mitochondria and oxygen (the final electron acceptor, highlighted in yellow) to form water. With no oxygen, water can't be formed, and the electron transport chain stops!!

This brings us to an interesting bit of history. The electron transport chain uncoupler, dinitrophenol (DNP), was first introduced as a pesticide because it bound protons in the space between the inner and outer mitochondrial membrane and brought them back into the mitochondria without going through the revolving door enzyme, ATP synthase (so no ATP is generated). This killed troublesome insects because they could not generate enough energy (ATP) to survive. Because of this property, DNP was then introduced in 1933 as a human weight loss drug but was pulled off the market in 1938 because it increased metabolic rate and body temperature and led to organ failure. However, it is still sold illegally today for weight loss.

Of course, we don't just get ATP from breaking down glucose, but from other foods as well! In **Fig 19A** I indicate where fats and proteins enter central metabolism to generate energy. For fats and proteins to enter any cell, they must first be broken down to their building blocks, fatty acids and amino acids, respectively. They are then taken up into cells by plasma membrane-embedded transporters and the fatty acids are broken down in the cell to acetyl-CoA and enter the TCA cycle. The 20 amino acids enter the TCA cycle at different points, with some entering after being converted to pyruvate.

What is important to know is that mitochondria need oxygen to "burn" the pyruvate to CO_2. So, the reason you and I, and all other eukaryotes (including plants) need to "breathe" is because of our mitochondria and the TCA cycle within them. But to be more specific, we need oxygen because it is needed for the electron transport chain to function. So, all I want you to remember is that without mitochondria and oxygen, we couldn't generate nearly as much ATP. As I will mention later in this chapter, the first life forms likely appeared in an atmosphere devoid of any oxygen and so they would not have been able to generate as much energy as eukaryotic cells today, nor could they use proteins or fats as an energy source.

Of note, while all eukaryotes and many prokaryotes have an electron transport chain, some primitive prokaryotes do not have the complete electron transport chain shown in **Fig 19B**. However, the pumping out of protons or other positively charged ions (like sodium) and their return into the cell is a common mechanism for generating ATP in all living organisms (36).

So, at your next cocktail party you can entertain the guests by informing them we only need to breathe in oxygen because we have mitochondria. And if there is still anyone who hasn't rolled their eyes and left, you can follow up with, "to be more specific, we could not generate any ATP in our mitochondria without oxygen because our electron transport pumps would stop." Maybe wait till people have had a few drinks and are too unsteady to get away.

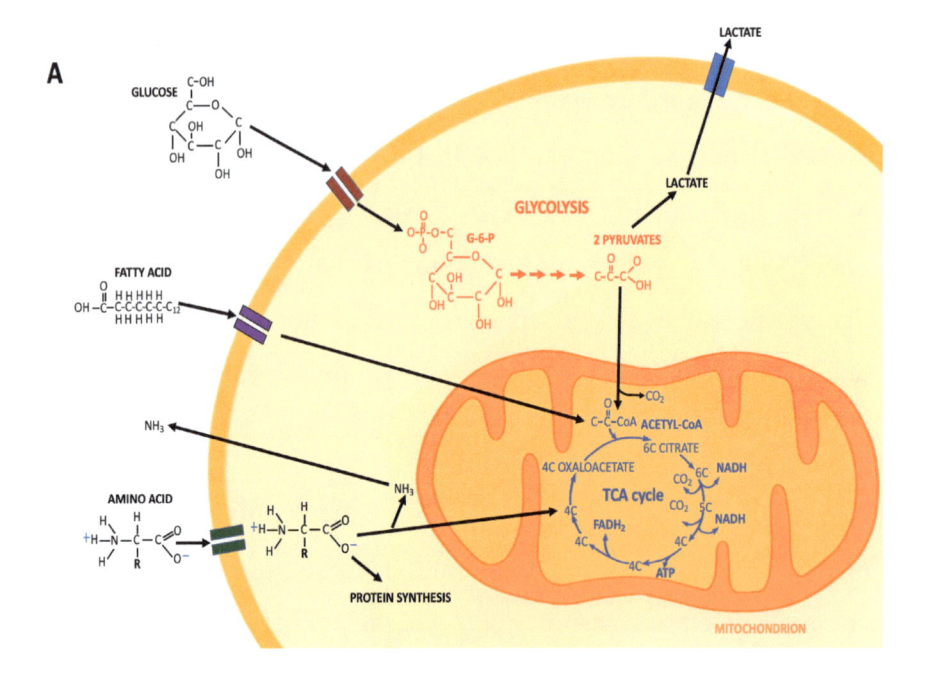

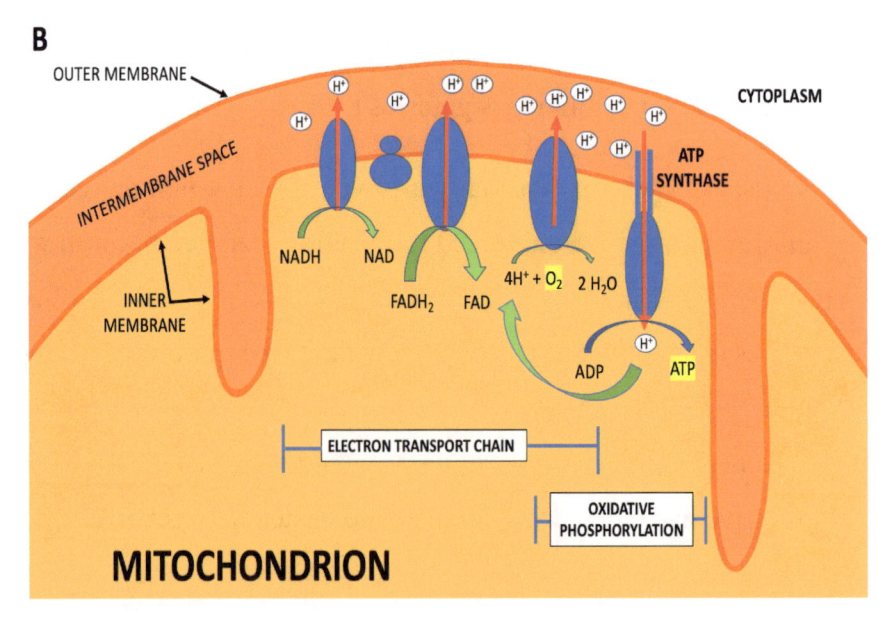

Fig 19. (A) Pathways in which starch (sugars), fat (fatty acids) and proteins (amino acids) are broken down for energy. Glycolysis occurs in the cytoplasm of both prokaryotic and

eukaryotic cells. The TCA cycle and reactions that occur in the electron transport chain take place within the mitochondria of eukaryotes. The only part of this breakdown that requires oxygen is the electron transport chain. **Proteins** are broken down to amino acids in the stomach and small intestine of vertebrates (animals with backbones). The amino acids are then taken up into cells via amino acid transporters embedded in the plasma membranes of all cells (37). Once taken up, they are either used to build proteins (anabolism) or they are broken down (catabolism). When broken down, amino acids lose their NH_3 group (as ammonia), which goes to the liver to be converted into urea and excreted. The rest of the amino acid (i.e., minus the NH_3 group) is broken down for energy within the TCA cycle. **Fatty acids** are broken down to acetyl groups and carried by CoA into the TCA cycle.

(B) The electron transport chain. This is a series of proteins embedded either in the inner mitochondrial membrane of eukaryotes or in the plasma membrane of prokaryotes. High-energy electrons generated in the TCA cycle are carried by NAD and FAD to a series of proteins which pump out protons. Specifically, NADH and $FADH_2$ release protons and high-energy electrons and the energy of the released electrons stimulates the protein pumps to pump out the protons. When the protons come back into the mitochondria to equalize the concentration of protons in and out, they pass through ATP synthase, which, in turn, generates ATP. The exhausted electrons combine with oxygen and the incoming protons to form water. While the electron transport chain of some prokaryotes (anaerobes) does not require oxygen, they still involve a proton or other positively charged ion (cation) pump to create a gradient across a membrane.

Time for a well-deserved break. That was a lot to digest (pun intended)!! I apologize for making you suffer through central metabolism, and I do not expect you to remember any of the details. They are not necessary to understand the rest of this book, but I wanted you to gain some appreciation, not only for the complexity involved in generating the energy needed for survival, but for the incredibly dedicated scientists who gave up countless hours of their lives to figure all this out.

Our Universe and How We Got Here

Part 2

In this second section we are going to talk about the role that DNA mutations play in evolution, a new way to look at the tree of life and how we humans classify all living things.

Our own genomes carry the story of evolution, written in DNA, the language of molecular genetics, and the narrative is unmistakable

Kenneth R. Miller

(e) DNA Mutations are Responsible for Evolution

Before we can discuss the 4-billion-year journey from our single celled ancestor to Homo sapiens (Latin for wise men...I don't think it counts if you are the one calling yourself wise), we first have to appreciate that while competition for energy is the major driver of evolution, DNA mutations (changes to the base sequence of DNA) allow evolution to happen. But before I go any further, I first have to explain the relationship between DNA mutations and evolution. If you are a single-celled organism, it is easy. Any mutations you acquire, if not repaired, will be passed on to your progeny when you divide into two daughter cells. However, if you are a multicellular organism like us, only mutations in your germ cells (sperm or eggs) will be passed on to the next generation. Mutations to the non-germ cells of our body, i.e., our somatic cells, which make up most of the cells in our body, are not passed on to our progeny and may, instead, lead to cancer.

OK. Let's now talk about the different kinds of mutations living organisms can get. Mutations can be a simple exchange of one base for another (the most common), the addition or deletion of one or more bases, or the movement (translocation) of a piece of DNA from one place to another in an organism's DNA. These changes occur because of imperfect DNA replication (copying) and/or DNA damaging agents and the imperfect repair systems that attempt to correct DNA damage.

In terms of imperfect replication, the enzyme responsible for duplicating DNA just before a cell divides to become two cells is called DNA polymerase. It typically puts in a wrong base by chance during duplication about once every one million bases (but this varies with each species because of slight differences in the amino acid sequence of their DNA polymerase). There are also DNA repair enzymes that are like little soldiers marching up and down the DNA looking for mistakes, which appear as bulges or narrowings of the DNA double helix "train track". They repair these mistakes by cutting out the wrong bases on the newly forming strand and putting in the right ones (see **Fig 20**). The faster a cell divides, the less time the DNA repair enzymes have to fix these bulges and narrowings, so faster replicating cells are more prone to mutations. Importantly, once the wrong base is put in and the cell divides, the wrong base can become a permanent fixture, as shown in **Fig 20**. Increasing this low but steady intrinsic (meaning from within) mutation rate is the effect of DNA mutating agents (extrinsic factors, meaning from outside), including ultraviolet light, cosmic rays and chemical carcinogens, which can overwhelm the DNA repair system and lead to increased mutation rates. Prokaryotes typically have higher mutation rates than eukaryotes, in part because they divide more quickly. For example, most bacteria can replicate every 20 - 30 minutes if there is plenty of food. Human cells, on the other hand, take from 8 to 24 hours to divide into two cells. This, however, is only true for those human cells that divide. Most human cells become incapable of dividing once they become mature, functioning cells. Further contributing to the faster mutation rate of prokaryotes is their lack of protective proteins (histones) covering their DNA (so they are more susceptible to UV-induced DNA damage).

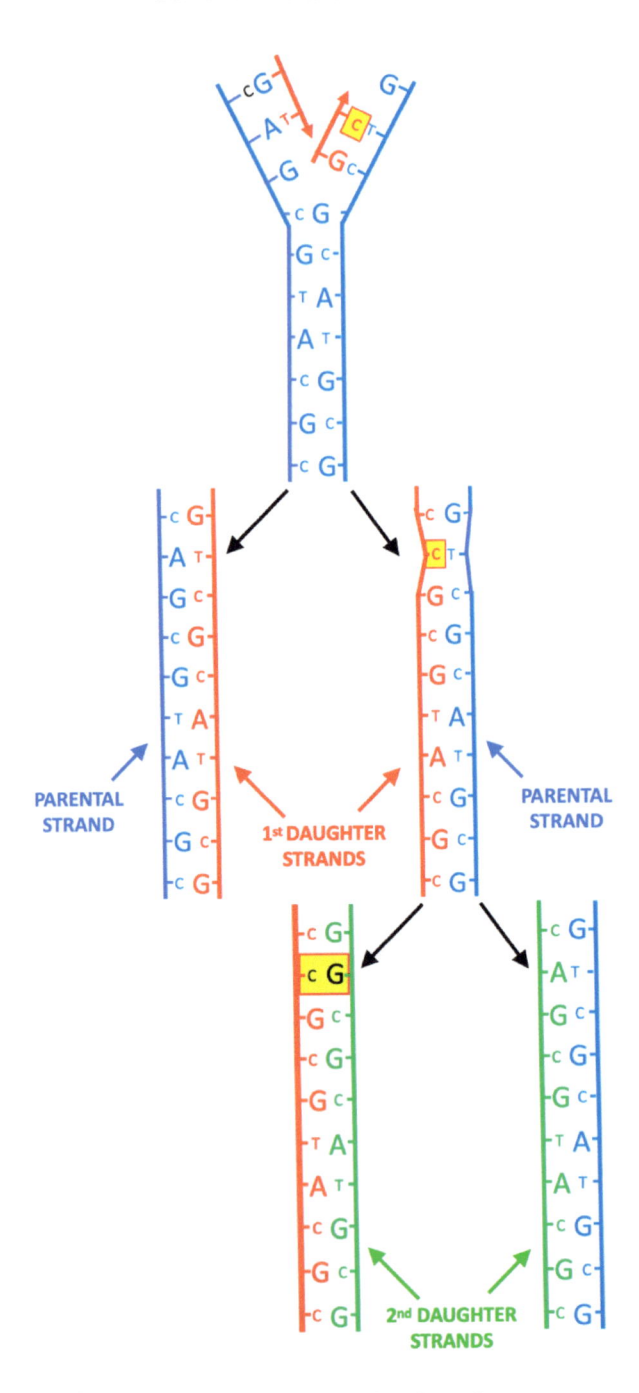

Fig 20. DNA Mutations can be repaired or become permanent.
Before a cell can divide to become 2 cells, the DNA must be

replicated (duplicated). To do this, the two "parental" strands (shown in blue) of the DNA open up like a zipper, as shown in the top-most panel, via breaking of the weak hydrogen bonds that hold the complementary bases together. New nucleotides come in to form complementary "daughter "strands. By mistake, C comes in, instead of an A to pair with a T. Since the C and T are both small bases this will narrow the new train track as shown in the right-hand side DNA in the middle panel. One daughter cell will get normal DNA (on the left) and the other (on the right) will get this mutant form of DNA. If this mutant DNA is not repaired before the daughter cell divides (by DNA repair enzymes, which use the parental strand as a template to put the correct base back in) it will become permanent, as shown in the left bottom panel. Specifically, you will get this new C=T opening up and a G will be brought in to pair with the incorrect C. So now you have a G≡C pair where you used to have an A=T, but the train track looks normal and so won't be repaired. Thus, you will permanently have a mutant cell that can keep dividing to go on to develop into a tumor.

It is important to remember that DNA mutations occur randomly. If a DNA mutation results in a simple base change (as shown in **Fig 20**) within a gene (a section of DNA coding for a protein), it can result in (a) no change to the amino acid sequence of the protein (given the redundancy of the codons), (b) a switch from one amino acid to a similar one, with no significant change to protein function or (c) a switch from one amino acid to a completely different one. This last scenario usually leads to a loss of function of the protein and is considered a harmful (deleterious) mutation. However, on rare occasions the amino acid change results in a more efficient enzyme and is considered an advantageous mutation. One way of decreasing the impact of deleterious mutations on an organism is to duplicate genes. This has happened quite often during evolution as a result of DNA replication errors and leads to two beneficial effects. First of all, if a deleterious mutation occurs in one of the two copies of a gene, the non-mutated gene can still code for a functional protein, which means the mutation is not detrimental/lethal to the organism. Secondly, by having two copies of a gene, mutations can happen to one of the genes (since there is no pressure on it to be functional) and it can evolve into a gene coding for a protein with a new function, increasing the repertoire of genes in an organism (which is advantageous).

I apologize if all of this sounds a bit Greek but all I really want you to remember is that DNA mutations occur randomly, and the price of evolution is deleterious mutations. However, these can be ameliorated (made less harmful) to some extent by gene duplications. Taken together, mutations lead to tremendous diversity and this diversity can allow some of the offspring to better survive under changing conditions.

(f) Tree of Life

Typically, when people think about evolution, they imagine something like the tree shown in **Fig 21A,** with us somewhere in the top branches. While it is certainly true that we are more complex than bacteria, it should be remembered that evolution is an ongoing process and all organisms, including bacteria, are changing constantly to adapt to changing environmental conditions. This means that **every** living organism alive today is the result of 4 billion years of survival. We are all survivors of an incredibly difficult 4-billion-year journey. To get more of a feel for what evolution is really like, I have plotted evolution over time and shown the major contractions and expansions in the number of different species due to natural disasters and competition (**Fig 21B**).

A

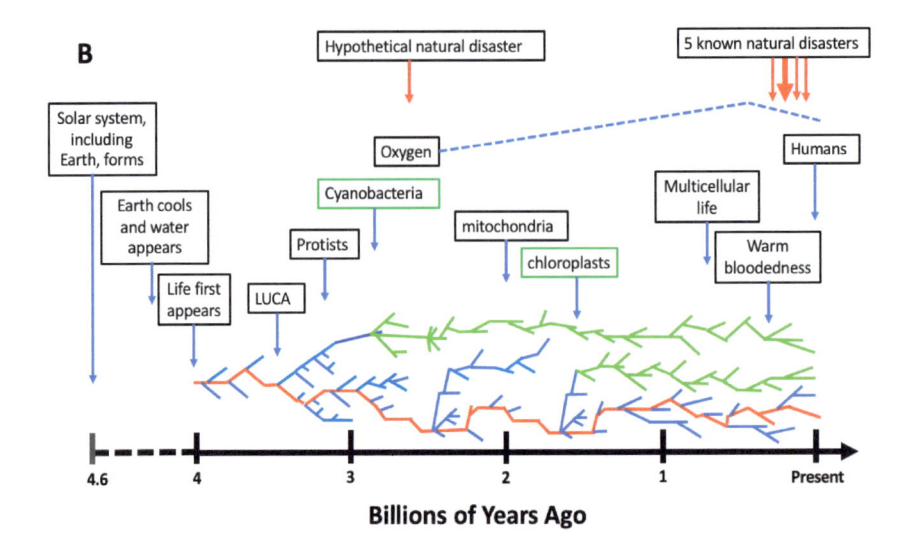

Fig 21. Evolution. (A) Classic evolutionary tree. (B) Current understanding of evolution over time. Each blue (non-chlorophyll-containing) and green (chlorophyll-containing) line indicates a unique species but, in reality, there were far more species existing at any one time than shown. At present, it is thought there are anywhere from 2 to 50 million species of living organisms, with single-celled organisms far outnumbering multicellular species. Amongst all the multicellular species, it is thought that there are currently over 307,000 different species of plants and over 66,000 vertebrates. The red line traces **our** own tortured ancestry. As indicated in the figure, our last universal common ancestor (LUCA), from which all existing life originates, is thought to have appeared around 3.5 billion years ago. Going forward in time, I've included what I consider to be the most critical steps leading from single-celled prokaryotes to us, such as the appearance of protists (single-celled eukaryotes), the oxygenation of Earth, the acquisition of mitochondria, and when multicellular life forms began. I will go into more detail on each step in Part 3 of this chapter. I have also included the 5 known major natural disasters that, at different times, almost wiped-out life on Earth, with the thickest red line representing the Permian-Triassic disaster that wiped out 90-95% of all species 250

million years ago. Based upon the frequency of these more recent, known disasters, we can assume there were many that occurred before them. I have added one hypothetical natural disaster earlier on in our evolution to illustrate this point. Important to remember is that after these great death events, there is lots of food (dead organisms) and very little competition for it. This likely explains why there is usually a great blossoming of species after natural disasters. With time, however, as numbers of living things increase, competition begins and many of the species get winnowed down. It is thought that at least 99.9% of all species that ever lived are now extinct.

(g) Dear King Phillip Came Over for Good Soup

I bet you didn't expect this title! It is a helpful mnemonic to remember how we classify all living things. When you look at the little red line in **Fig 21B**, which traces our specific ancestry from a common single-celled prokaryote, it is helpful to remember (a) that the earlier we split off from a common ancestor, the more different we are likely to be from the other descendants of that common ancestor and (b) every living thing today, including us, is a survivor of 4 billion years of evolution (so give yourself a pat on the back!).

Although we humans may be overly preoccupied with trying to compartmentalize the chaos around us, our particular obsession with classifying living things is actually helpful, in part because it gives us insight into evolutionary relationships. For example, the complete description of Homo sapiens is:

Dear = Domain, of which there are three: Eukaryotes, Bacteria and Archaea. Humans are **Eukaryotes**, since every one of the 30 trillion cells in our body contains a nucleus. Actually, I lie. Every cell except our mature red blood cells, since, as these cells mature, they extrude (toss out) their nucleus to make more room for the oxygen carrying protein, hemoglobin.

King = Kingdom, with humans being **animals**, rather than plants, fungi, protists, bacteria or archaea.

Phillip = Phyllum, with humans being **chordates** (vertebrates, which means animals with a backbone), rather than arthropods, annelids, moluscs or nematodes.

Came = Class, with humans being **mammals** (which means the young are nourished by milk from a mother's mammary glands), rather than amphibians, reptiles, fish or birds.

Over = Order, with humans being **primates** (mammals adapted to living in trees, with feet and hands that can grasp) rather than the 19-21 other mammalian orders.

For = Family, with humans being **hominids** (great apes), rather than lemurs, lorises, tarsiers, New World or Old-World monkeys.

Good = Genus, with humans being **homo** (latin for man), rather than the extinct Australopithecus, Homo erectus or Homo Neanderthals.

Soup = Species, with humans being Homo **sapiens** (wise men, which, on reflection, may be a misnomer).

To make life easier we typically only use Genus and species (in Latin, of course) to describe all living things. This was first popularized by the highly acclaimed Swedish botanist, Carl Linnaeus, in the 1700s. Time for a break!!!

Our Universe and How We Got Here

Part 3

In this section we are going to discuss the possible earliest life form(s) on Earth and some of the critical steps in evolution that led to us, the immodest Homo sapiens.

It is not the strongest of the species that survives nor the most intelligent. It is the one that is most adaptable to change

Charles Darwin

(h) The Earth Cools and Life Begins

It is thought that our sun and all the planets in our solar system, including Earth, formed about 4.6 billion years ago, and that life started on Earth soon after, when our planet started to cool (about 4 billion years ago). It could not have started sooner because the Earth was too hot to have liquid water. Interestingly, although water currently covers about 70% of the Earth's surface, we are still not sure if this water came primarily from comets ("dirty snowballs" consisting mainly of ice that leave visible water vapor tails when they orbit close to the sun), asteroids (rocks which have hydrogen and oxygen trapped inside their minerals) or from the Earth itself (which, like the other planets in our solar system, formed from the gas and dust left over from the formation of the sun (the solar nebula)). Recent data suggest that it may have indeed come from the Earth itself (38).

(i) Critical Steps in our Evolution

Originally, I was going to call this section "the 10 Critical Steps in our Evolution", which of course would have been a totally arbitrary number. It brought me back to when I was an undergraduate student at McGill University in the 1960s. I was writing the final exam in a second-year botany course and the final question was a gift, "Pose a botanical question and answer it". I wanted to show off (I had studied really hard) so I wrote "Name the 20 characteristics of ferns". Anyway, under the pressure of the exam I could only remember 19 so, just before we had to hand in our test papers, I scratched out my original question and, underneath, wrote "Name 19 of the 20 characteristics of ferns". I imagine my professor must have smiled when he saw my two titles, but I never found out how I did on that question. So, although my current evolutionary list is far from exhaustive and somewhat arbitrary, I have tried to home in on what I consider the most critical steps leading from simple single-celled prokaryotes to us. I've included these critical steps in **Fig 21.** The possible origin of the first single-celled prokaryotes will be addressed in Chapter 5.

3.5 to 4 billion years ago - Chemotrophs

Life probably first appeared when the atmosphere was composed primarily of methane, spewed from volcanoes, both above ground and at the bottom of the oceans. Volcanoes are simply ruptures in the crust that allow hot lava and gases to escape from the semi-solid mantle within the Earth. Since the core of our planet was likely in greater turmoil 3.5 to 4 billion years ago than it is today, there were probably many more volcanoes back then, most of them at the bottom of the seas. In the late 1970s, scientists were shocked to find single-celled microorganisms living in total darkness on the sea floor beside such volcanoes. Specifically, they were living beside hydrothermal (hot water) vents. These hydrothermal vents are generated by seawater seeping into deep cracks (fissures) in the sea floor, often between two tectonic plates. The water gets superheated from the underlying magma and comes spewing back up filled with high concentrations of hydrogen, hydrogen sulfide, carbon dioxide, ammonia and methane from the magma. The scientists found that these microorganisms were thriving at temperatures in excess of 100°C (212°F), by using the vent-generated hydrogen, hydrogen sulfide and methane (natural gas) as an energy source. Specifically, they got energy from these simple molecules via oxidation. So what does that mean? You are probably familiar with the idea of oxygen reacting with metals, causing them to rust (becoming oxides). In chemistry and biology, we use a broader definition of oxidation, which not only includes a molecule binding oxygen (leading metals to rust) but also a molecule losing electrons. In this definition, the oxidation of a molecule (the loss of electrons) is always accompanied by the reduction, or gain of electrons, by another molecule (oxygen loves electrons). So, you are basically getting energy by transferring electrons from these simple chemicals to electron acceptors within the cell and this causes a release of energy that is used to generate ATP. I touched on this earlier when I talked about the amazing electron transport chain and some version of this chain being present in all living things. Chemoautotrophs that are alive today then use this ATP they make from oxidation/reduction reactions to convert dissolved carbon dioxide (CO_2) into organic molecules like glucose and amino acids, etc.

Today, the presence of these heat loving microorganisms (thermophiles) at the bottom of the sea makes possible a thriving ecosystem composed of snails, clams, mussels, etc., that feed on them. To avoid being eaten, some of these single-celled thermophiles have formed symbiotic (mutually beneficial) relationships with giant tubeworms and giant white clams. The giant tubeworms, which are a hallmark of hydrothermal vents, are especially fascinating. As adults they have no mouth or digestive system and can only get food from the millions of thermophiles they house inside them. They are so well nourished by these microorganisms that they are the fastest growing invertebrates (organisms with no backbone) on Earth, growing up to 2 meters (6.6 feet) in a single year!

Just for fun, let's imagine you are the first living organism on Earth. You are likely a single-celled bacterial- or archaeal-like cell, perhaps close to a hydrothermal vent 3.5 to 3.9 billion years ago (39). Now, as the first living organism on Earth you can't get energy from eating other living or dead organisms since neither exist yet. However, there may have been some amino acids already in existence from abiotic sources (i.e., not derived from living or once living organisms) that could have been used as an energy source. I will touch on this more in Chapter 5. As the first living organism, you likely do not have any photoresponsive molecules to get energy from the sun and you couldn't use them even if you had them since you are at the pitch-black bottom of the ocean. So, you likely are getting energy to stay alive by taking up these simple chemicals dissolved in the boiling water. As mentioned above, this makes you a chemoautotroph, which means you can make ATP from simple chemicals like hydrogen, hydrogen sulfide, methane or ammonia (35).

So, a possible common ancestor of all currently living things may have been a single cell resembling modern day archaea, capable of using hydrogen, hydrogen sulfide, methane or ammonia as an energy source and thriving in an atmosphere free of oxygen (a chemoautotroph). As shown in **Fig 22**, this theoretical common ancestor likely had a cell wall, a circular DNA, a plasmid or two, ribosomes and many different proteins, like bacteria and archaea today. This does not mean that the first life forms (i.e., those capable of reproducing themselves) on Earth had all these characteristics (and this will be discussed in the next chapter) but that the first **common ancestor,** also called the last universal common ancestor **(LUCA)** that has given rise to all current life forms may have had them **(Fig 21)**. Why do I include plasmids? They are small circular

DNAs that can replicate independently of their "host" DNA and are present today in bacteria and sometimes in archaea and eukaryotes as well. They only carry 1 or 2 genes, but these genes can benefit the survival of the hosts, such as endowing them with antibiotic resistance. Importantly, plasmids can be given to neighboring microorganisms by direct cell contact, to increase the survival of these neighbors. These neighbors do not necessarily need to be of the same species to exchange plasmids. Like viruses, plasmids are not generally considered alive and I am mentioning them for two reasons. One is to illustrate that evolution is not always driven by competition but sometimes by **collaboration** and two different species might work together to ensure they both survive, in this case via exchanging their plasmids to increase survival. The other reason I am mentioning plasmids is because they may be the forerunners of all viruses that exist today. I say "may be", because viruses may also be degenerated forms of primitive cells. Either way, viruses could not have existed before cells because they need cells to reproduce!

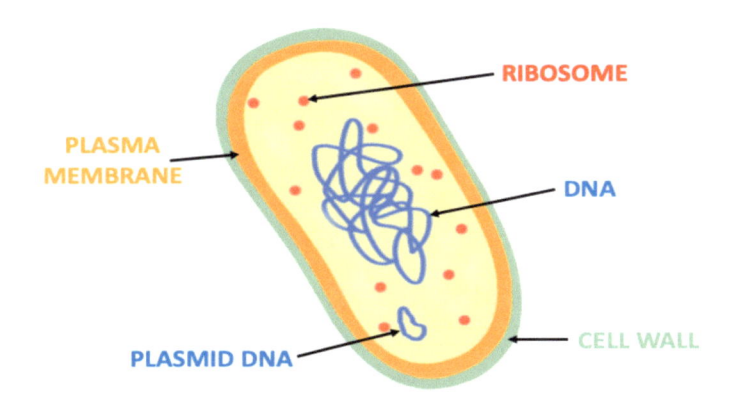

Fig 22. A hypothetical common ancestor of all living things today. This prokaryotic cell (i.e., with no nucleus or organelles) may have been a chemoautotrophic thermophile (an archaea-like cell that can satisfy all its energy needs from simple chemicals and thrive in boiling hot water). I am hypothesizing the presence of a cell wall or outer cell membrane (green) because the electron transport chain pushes protons into the space (not shown) between the plasma membrane (orange) and cell wall, where they pile up and come back into the cell to generate ATP. Without a cell wall or outer cell membrane, the protons would just float away and not be concentrated. I am also proposing the presence of a plasmid (small dark blue circle) and ribosomes (red dots). The large folded up large dark blue circle is the cell's DNA.

3.5 billion years ago - Cyanobacteria/Chlorophyll: Life Changes the Atmosphere

We are alive today because of cyanobacteria (also called blue-green algae)! Let me repeat that. We, and all other oxygen-breathing life forms, are alive because of cyanobacteria (see **Fig 21**). If life indeed first started beside hydrothermal vents at the bottom of the ocean (and not in "warm little ponds" as suggested by Darwin (40)), some of these life forms must have mutated so they could live at or near the surface of the ocean in order to give rise to cyanobacteria. These bacteria use sunlight to split water (H_2O) into hydrogen, which they keep to generate energy, and oxygen (O_2), which they release. While other bacteria today can use photosysnthesis to generate energy, cyanobacteria are the only bacteria that can produce oxygen as a byproduct of photosynthesis (similar to all plants). This ancient prokaryotic family was likely the first to contain chlorophyll, the green pigment responsible for photosynthesis. During photosynthesis, sunlight causes an electron in chlorophyll to gain enough energy to leave the chlorophyll molecule and enter the electron transport chain to generate ATP. The chlorophyll with its missing electron gets an electron back from water. Specifically, to get this electron from water, the water first has to get split into hydrogen and oxygen. The hydrogen gives its electron to chlorophyll and its proton gets pushed into the space between the cell membrane and cell wall in the electron transport chain. The oxygen released when water is split goes into the surrounding water as a byproduct of photosynthesis. The cyanobacteria use the ATP they just made in the electron transport chain to convert water and carbon dioxide (CO_2, dissolved in the ocean) into the simple sugar, glucose. This glucose, in turn, generates ATP to keep cyanobacteria alive. As an aside, it is worth mentioning that cyanobacteria have internal gas vacuoles which they use to regulate their buoyancy, allowing them to move up to get sunlight and down to avoid harsh UV rays.

Evidence for very early life at the Earth's surface include fossilized stromatolites (stony structures), found in 3.5-billion-year-old rocks (41). These stromatolites, which still exist today, are made up of layers of mucilage-secreting microorganisms, including cyanobacteria, with the secreted mucilage adhering to sand to form protective mats.

OK. So cyanobacteria likely first appeared some 3.5 billion years ago close to the surface of our early oceans and started pumping out oxygen into them. Since the early oceans were full of dissolved iron (Fe, the most common element on Earth and likely spewed into the oceans from volcanoes), this oxygen first reacted with the Fe in the oceans to form water insoluble iron oxide (FeO_3). This iron oxide precipitated (fell) to the bottom of the ocean and turned Earth into a red planet. Thus, there may have been up to 1 billion years between the first appearance of cyanobacteria (fossil records suggest 3.5 billion years ago) and any oxygen in the atmosphere (2.4 to 2.9 billion years ago, **Fig 21**) because the ocean "absorbed" the increase in oxygen for many years by binding it to dissolved iron (42).

Once oxygen made it into the atmosphere, it led to the formation of the ozone layer (since ozone is derived from oxygen). The formation of this ozone layer protected life floating at or near the ocean surfaces from the DNA damaging effects of cosmic rays. As well, this atmospheric oxygen likely reacted with the greenhouse gas, methane, depleting it and causing Earth's first glaciation, turning Earth into a giant snowball (43). Lastly, and most importantly, the appearance of oxygen in the oceans and atmosphere led to the near extinction of oxygen intolerant (anaerobic) organisms like Archaea (and this period is sometimes called the Oxygen Catastrophe). Their death provided lots of food for aerobic (oxygen loving) bacteria to multiply. The few archaea that survived, fled to niches devoid of oxygen like the ocean floor, where the dissolved oxygen levels are very low, and elsewhere, like our colon (large intestine). A fun fact is that only archaea in our colon can generate methane gas. However, while this leads to noisy farts it does not lead to smelly farts. Those come from bacteria and archaea in our colon that produce hydrogen sulfide!

Today, cyanobacteria are aquatic prokaryotes, mostly unicellular (but can form multicellular filaments), that often bloom when there are lots of nutrients in the water. Their blooming leads to the familiar scum we see on the surfaces of ponds. Cyanobacteria also live symbiotically with fungi to form lichen. Within the lichen, the cyanobacteria get protection while the fungus gets energy from the cyanobacteria's photosynthesis. This kind of symbiosis is called ectosymbiosis because one species lives on the outside surface of the other (not inside as will be mentioned shortly). Lichen played a critical role in enabling life to migrate from the water to the land because lichen can break up bare rock to form soil. This allowed plants to subsequently gain a foothold on the land.

1 to 2 billion years ago – The appearance of Single-Celled Eukaryotes and Endosymbiosis

The presence of all living things today did not result just from nasty, kill-or-be-killed competition but also from collaborations at different times in our evolution. For example, the first appearance of larger, single-celled eukaryotes (protists, **Fig 21**) with their well-defined nuclei and organelles, may have been the result of partnerships amongst many different types of prokaryotes in underwater microbial mats. Within these dense mats, cells at the surface may have been able to use the rising oxygen levels to "burn" sugars, amino acids and fatty acids all the way to carbon dioxide for maximal energy production. The cells deeper within the mats, out of reach of the newly generated oxygen, could not perform these same metabolic reactions, but instead used the wastes produced by the top layer of cells to survive. In other words, the wastes of one group of cells were acting as the fuel for another. Over time, these metabolic partnerships may have become more intertwined, with one cell type merging with another, creating a new type of cell with more cellular complexity. These more complex cells evolved into protists, the first single-celled eukaryotes (44).

There were two cases of a special kind of symbiosis, where one organism swallowed (engulfed) another that dramatically changed the world. Specifically, this engulfment led to what we now know as **mitochondria** in all eukaryotes, and **chloroplasts** in all plants (**Fig 21**). This kind of symbiosis is called endosymbiosis (endo meaning internal or inside of the organism).

Mitochondria

Shortly after the introduction of oxygen into the atmosphere by cyanobacteria, there was a dramatic increase in the number of different bacterial species on our planet, in part because there was little competition amongst them with all the food that became available (dead archaea). When oxygen was first displacing the methane in the atmosphere (which was generated not just by volcanic eruptions but also by methane spewing archaea) one of these newly evolved bacterial species acquired the ability to use this oxygen to "burn" glucose all the way down to carbon dioxide, unlike its brethren who were only capable of breaking it down to lactic acid or ethanol. This new development resulted in 1 glucose molecule yielding up to 38 molecules of ATP instead of just 2!! This gave this "super-charged" bacterium, and all its offspring, a tremendous survival advantage since it could survive on far less food.

Now here comes the critical event in our evolution. One of these "super-charged" bacteria was eaten/engulfed by a larger cell (likely a eukaryote). This engulfment of smaller bacteria by larger cells was not unusual and is one of the reasons that getting big is so evolutionarily advantageous. However, the engulfed bacterium managed to prevent its own digestion, which is also not an unusual phenomenon and still occurs today. For example, when the bacterium, Mycobacterium tuberculosis, which causes tuberculosis, is engulfed by a macrophage in our lungs, the mycobacterium inhibits its own digestion and thrives in this new, protected environment. So too, back then, this super-charged prokaryote (which may have been an ancestor of what we call rickettsia, i.e., bacteria that need to enter the cytoplasm of eukaryotic cells to replicate) thrived in its protected environment and gave some of the

ATP it generated to its new host, enabling the host to survive with much less food. This gave the host such an incredible survival advantage that all eukaryotes alive today, from single-celled eukaryotes (protists) to all animals and plants, house these bacteria, now known as mitochondria. These mitochondria have their own circular DNA and ribosomes. They replicate independently from the host DNA and they share many characteristics with present day bacteria. Humans can have up to 2,500 mitochondria in a single cell (cells that need more energy like muscle cells tend to have more mitochondria). This endosymbiosis event could very well be one of the biggest singular steps in our evolution since the higher ATP levels that could be attained from the acquisition of mitochondria may have made multicellular life possible. However, it is important to remember that these mitochondria, in all of our cells, need oxygen to "burn" glucose all the way to CO_2. This is called aerobic metabolism (since it needs oxygen), and while it provides much more energy than anerobic metabolism, some think the acquisition of these highly efficient bacteria might have been a "deal with the devil" or at least a "two-edged sword". This is because we now cannot live without oxygen and, more importantly, mitochondria generate reactive oxygen species (ROS) as a side product when they burn food to CO_2 (which we breathe out). ROS can damage our DNA and lead to aging and cancer. Also, when cells in our body die, we sometimes release mitochondria or mitochondrial debris from our dying cells into the surrounding extracellular milieu. Sometimes our immune system mistakes this released mitochondria as invading bacteria and overreacts. This overreaction might lead to chronic inflammation and some autoimmune diseases.

Chloroplasts

Another example of endosymbiosis that must be mentioned involves our old friend, cyanobacteria. There is evidence that plant and algae chloroplasts (responsible for the green color and photosynthetic abilities of plants and algae) were originally cyanobacteria, that got engulfed by an ancient single-celled ancestor of current plants and algae. Therefore, all the plants and algae on Earth can thank their very existence to cyanobacteria! Chloroplasts absorb energy directly from the sun to convert CO_2 into glucose (photosynthesis). They can then break down some of this glucose to generate ATP. An important byproduct of photosynthesis is oxygen. Oxygen is expelled by plants and both plants and animals breathe it in to break sugars, fats and proteins down to CO_2 in mitochondria (present in both animal and plant cells) to generate ATP (see **Fig 19**). Similar to mitochondria, chloroplasts have their own ribosomes and a circular DNA, and the DNA sequence of chloroplasts is very similar to that of cyanobacteria! Since chloroplasts are present in plants and algae and not animals, whereas mitochondria are in all eukaryotes, we think the incorporation of mitochondria occurred before chloroplasts during evolution (see **Figs 21 & 23**).

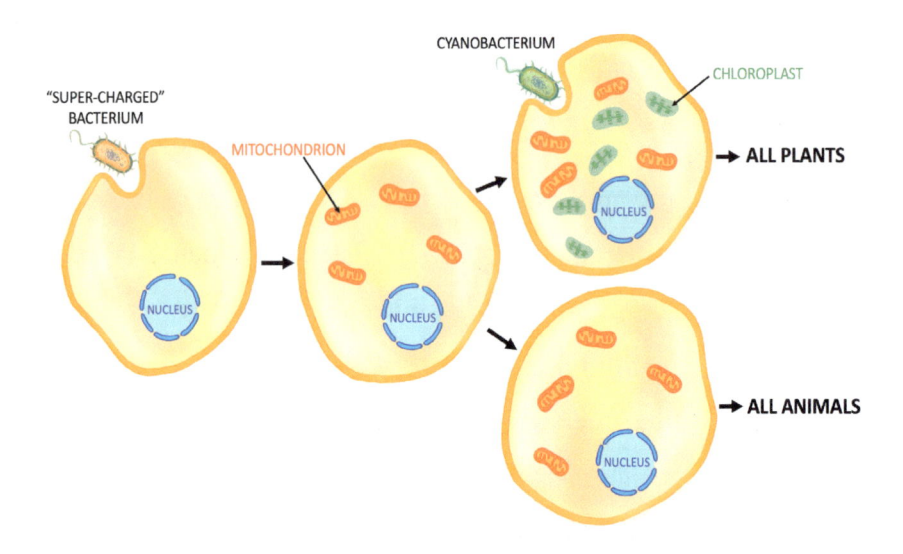

Fig 23. The introduction of mitochondria and chloroplasts during evolution. Around two billion years ago, a large cell engulfed a "super-charged" bacterium capable of aerobic respiration. This bacterium avoided being digested and, over time, evolved into a mitochondrion: an ATP generating organelle present in all eukaryotic cells. Similarly, a single-celled common ancestor of plants and algae engulfed a cyanobacterium, capable of photosynthesis. This cyanobacterium avoided digestion and evolved into a chloroplast. Because mitochondria are present in both plants and animals whereas chloroplasts are only present within plants, it is likely that mitochondria appeared chloroplasts. However, it is conceivable, though unlikely, that chloroplasts were incorporated before mitochondria and then subsequently lost in the ancestors of modern-day animals.

1.5 to 2 billion years ago – The Sexual Revolution

The introduction of sexual reproduction is thought to have allowed for much more genetic variation than was attainable by asexual reproduction, which typically occurs via budding, spore formation or binary fission (splitting in half to form 2 daughter cells) (see **Fig 24A**). Metaphysically speaking (which means looking beyond physics at questions that cannot be answered through current scientific experimentation, like what is reality and consciousness), one could say prokaryotes are immortal since they just keep splitting in half. But that brings up the concept of consciousness (self-awareness) and when in our evolution it appeared, which is beyond the scope of this book.

Sexual reproduction occurs in almost all multicellular organisms and can also occur in single-celled protists (remember, these are eukaryotic single cells with well-defined nuclei, like amoeba) under stressful conditions (45). While asexual reproduction typically results in more offspring than sexual reproduction (since all asexual organisms can give rise to offspring rather than just the females), all the offspring are genetically identical to the parent, barring mutations. The early stirrings of sexual reproduction likely began in prokaryotes when bacteria began exchanging genes, via plasmids. In eukaryotes, this became full blown sex, in which individual offspring inherited a random, mixed bag of a mother's and father's genes. This is explained in more detail in **Fig 24B**. This random inheritance increased diversity, and consequently increased the chances of a species survival since some offspring could adapt better to changing environments. This diversity also made speciation possible, in which one species branches off into two separate species. Speciation usually occurs when the offspring of one species end up living in two very different environments, with different external pressures. The other side of the coin, of course, is that sexual reproduction could also eliminate defective genes from the gene pool via death at any stage before maturity/mating. Perhaps most importantly, on the positive side, it led to mate selection, where males and females could vie for mates with traits that might maximize offspring survival.

Using humans as an example, every "somatic" cell in **our** body, meaning every cell except our gametes (i.e., sperm or eggs), is **diploid** (i.e., it has two copies of all our 20,550 genes, scattered over the 23 chromosomes from our mother and the 23 from our father). Because of this, we cannot simply produce offspring by fusing one of our somatic cells, with its 46 chromosomes, with one from our mate. This would lead to 4 copies of every gene and 92 chromosomes. Moreover, with every generation, this copy number would double and we would quickly end up with too much DNA for our nuclei to house, and require a wasteful amount of energy to replicate. Thus, to get around this problem, we generate **haploid** eggs and sperm (with only one copy of every gene) through a process called meiosis. (**Fig 24B**).

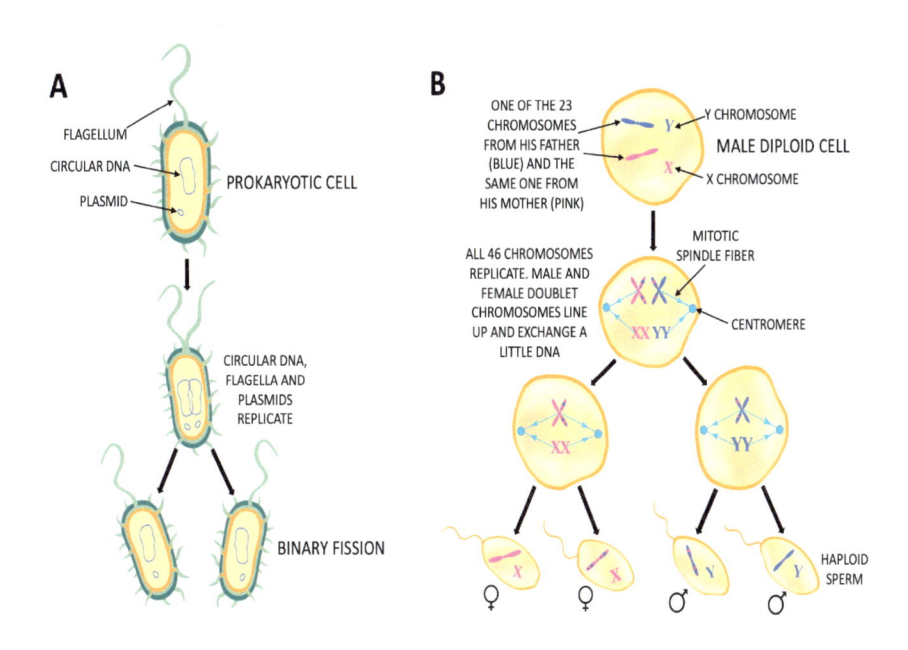

Fig 24. Asexual versus Sexual Reproduction. (A) Prokaryotes, reproduce asexually by making a copy of (replicating) their circular DNA (mitosis) and then splitting in half (binary fission). They are also haploid, which means they have only one copy of their approximate 2000 genes (this number varies considerably with the type of bacteria or archaea). **(B)** Eukaryotes, especially multicellular ones, reproduce sexually and are typically diploid, meaning they have two copies of each gene (i.e., two copies of the 20,500 genes in humans, but the total number of genes in eukaryotes varies from one species to another). Chromosomes in multicellular organisms are typically linear and segmented (i.e., divided up into several chromosomes rather than having just one big chromosome, and this segmentation allows for more diversity). Shown at the top in (B) is a normal human **male** diploid cell, which has 23 chromosomes from his mother and 23 from his father. For simplicity, I am only showing 2 chromosomes from his father (in blue) and 2 from his mother (in pink). Two of these chromosomes are sex chromosomes. Typically, females have two XX sex chromosomes while males have one X and one, much smaller, Y chromosome. In meiosis, all 46 chromosomes replicate and line up as doublets in the middle of the

cell at what is called the mitotic plate. Here, a specific male chromosome doublet lines up with its female counterpart (homolog) so, in the second cell from the top, you see 4 similar chromosomes. At this point there is some exchange of DNA between the male and female chromosomes to increase diversity in the subsequent sperm. The mitotic spindle fibers then pull the doublets to opposite poles with a random distribution of male and female chromosomes and cell division occurs. The 2 daughter cells then go on to divide once more (without replicating their DNA first) so you end up with an equal number of male (♂) and female (♀) sperm cells with only 1 copy of each chromosome, ready to fertilize egg cells that have gone through a very similar process.

As mentioned earlier, one major advantage of being diploid is if a mutation inactivates a gene, the other copy of this gene can rescue the lost function. Another major advantage, as shown in **Fig 24B** is that during the generation of germ cells (i.e., sperm and eggs) for reproduction (via meiosis), the genes obtained from the father are shuffled randomly with the genes of the mother so that each sperm (and each egg) is genetically unique and thus gives rise to offspring that are not only distinct from their parents, but also from each other. This is in sharp contrast to asexual reproduction in prokaryotes where the offspring are genetically identical to their parents (barring mutations). Thus, sexual reproduction results in much more genetic diversity, and those offspring that randomly inherit better survival traits for the specific environment they are in can be selected in future matings (natural selection).

Interestingly, the evolution of sexual reproduction has led to some rather bizarre life cycles, especially amongst parasites which use one or more species as hosts for asexual reproduction (secondary hosts) and different species for sexual reproduction (primary hosts). Several of these parasites appear to change the behavior of their hosts to increase their chances of survival. For example, *Toxoplasma gondii* is a single-celled eukaryote that sexually reproduces only in cats (its primary host). It is shed into the cat's feces and is eaten by mice and rats, or, via contaminated water, by humans (all secondary hosts), where it reproduces asexually. In mice and rats there is evidence this parasite goes to the rodents' brain and makes them more impulsive and reckless by, for example, causing them to lose their fear of cats and/or become attracted to the smell of cat urine. These behaviors increase the chances of these secondary hosts being eaten by cats. Intriguingly, about 30% of all humans are infected with *Toxoplasma gondii* and there is evidence it increases risk-taking, schizophrenia and suicides and may also cause cognitive decline in some susceptible, otherwise asymptomatic patients (46). This begs the question, does your friend really like bungee jumping, or are they just infected with *Toxoplasma gondii?*

550 million to 1 billion years ago – Flagella, Cilia, Eyespots and Chemical Warfare

We don't really have a good handle on when flagella (1 to 8 long, whip-like hairs present on the outside of bacteria, archaea and eukaryotic cells) or cilia (up to thousands of tiny hairs on the surface of eukaryotic cells) first appeared so I am taking serious liberties with this timeline, but there is strong evidence they have been around for at least 550 million years. In terms of cell movement, the biologically simplest system, and thus likely the first system to be developed, was a gas-filled vacuole in a cell that could increase or decrease in size to regulate movement up or down in the ocean. The far more complex flagella likely came after and then cilia even after that (since cilia are only present in eukaryotes today). The development of an eyespot (i.e., a pigment molecule that can discriminate between light and dark because it has an electron that can get excited at a specific wavelength of light) allowed coordination with flagella or cilia to take an organism to or away from the sunlight. The reason I wanted to mention these features is that they highlight an important theme of evolution; a constant arms race to detect and respond quickly to the environment. I also wanted to mention cilia because they serve to highlight a recurrent theme in evolution, i.e., repurposing. For example, cilia have been repurposed many times during evolution from its original purpose of unicellular movement, to sweeping in food (for example, in paramecia and sponges), to drawing water into the gills of fish (to extract the oxygen dissolved in water) and, still later, to pushing debris out of our lungs.

Speaking of an arms race, I don't know if chemical warfare started earlier than flagella, cilia and eyespots, but many microorganisms today pump out (secrete) metabolites that are toxic to other microorganisms. This enables the secretors to be more successful at proliferating and surviving than their competitors. The existence of this microbial, chemical warfare was revealed in 1928 when Alexander Fleming was a little sloppy during one of his experiments with bacteria. Specifically, he let a mold accidentally get into one of his bacterial plates and found it secreted something (penicillin) that prevented the bacteria from growing. This discovery led to the common use of antibiotics today, which has saved countless lives. An interesting historical note on this is that ancient Egyptians wrapped moldy bread around infected wounds, and so were aware of the ability of mold to save Egyptians from deadly infections (via secreting, unbeknownst to them, anti-bacterial substances). Unfortunately, this knowledge was lost over time.

730 to 550 million years ago – Multicellularity

It is generally thought that multicellular organisms first appeared during the Cambrian Explosion, which occurred about 550 million years ago (47) (**Fig 21**). However, the earliest multicellular animals (i.e., the sea-dwelling sponges, which are eukaryotes) may have appeared earlier than this, perhaps as a result of the increasing concentrations of oxygen in the oceans (before it increased in the atmosphere). As I mentioned earlier, eukaryotes have mitochondria and thus can use oxygen (produced by cyanobacteria) to generate a lot of ATP. This, in turn, may have provided sufficient energy to permit multicellularity. Also of note, there is evidence that single cells may have become multicellular more than once in evolution and then reverted back to single cells (48), so the exact timing of the first multicellular organism is still an open question.

Biochemically, the appearance of multicellular organisms corresponded with an increase in sticky molecules at the cell surface. For animal protists these sticky molecules might have been proteins like present day cadherins and integrins. For plant protists, because of their cell wall, they might have been molecules like the present-day soluble fiber, pectin.

Why multicellularity is so important is that it allowed, for the first time in evolution, cell specialization to take place within an organism. This specialization of cells within a multicellular organism occurred by the selective "turning on and off" of genes (epigenetic changes) within different cells of a multicellular organism. How did this occur if all the cells start out identical? To understand this phenomenon, it is important to first know that it is very common for single-celled organisms to secrete molecules, whether they be end-product metabolites the cells don't want, toxins to eliminate competitors or communication molecules to aid in mating and coordinated movement. Through evolution, these molecules were likely repurposed for cell specialization within multicellular organisms. For example, cells in the middle of a growing multicellular organism may be exposed to more of these secreted molecules than those at the periphery. As a result, a gradient of secreted molecules is generated. This gradient this can lead to distinct signaling cascades within different cells, and thus the turning on and off of different genes and the appearance of different proteins (**Fig 25A**). Another mechanism for generating different cell types at an early stage in the development of a multicellular organism is via random, unequal distribution of cell contents when cells divide (49) (**Fig 25B**). Together, these two mechanisms make possible the generation of specialized cells that, together, are superior to a single "Jack-of-all-trades" kind of cell trying to carry out all the functions needed to survive in a highly competitive world.

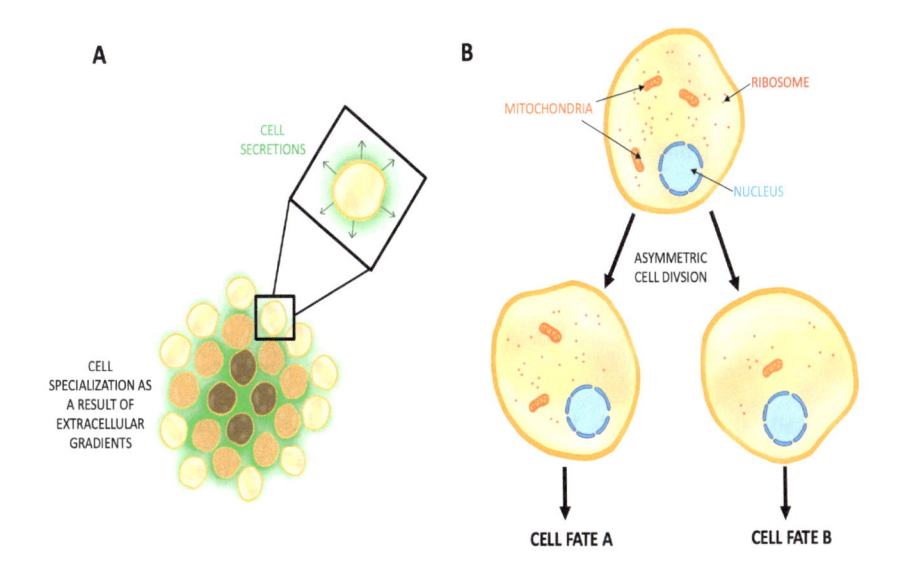

Fig 25. Extracellular gradients and unequal cell division make cell specialization within multicellular organisms possible. (A) The position of a cell within a multicellular organism affects the levels of secreted molecules to which it is exposed. The cells in the

center of a growing multicellular organism (shown in darker brown) are exposed to more secreted molecules than cells at the periphery (shown in lighter brown). **(B)** Unequal distribution of cytosolic contents and organelles during cell division can lead to different cell fates.

470 million years ago - Coming onto the Land

Coming onto the land was a huge step (there's a pun in there somewhere) in our long path to becoming humans. The first organisms to make this migration were likely plants and specifically plants similar to present day lichens. Lichens are one of the hardiest organisms on Earth, capable of surviving on bare rocks and withstanding harsh ultraviolet radiation and the dry conditions present on land. These "pioneer plants" are actually composed of two separate species, one being either a member of the large family of algae (chlorophyll-containing eukaryotes) or blue-green algae (our chlorophyll-containing prokaryote, cyanobacteria), encased within a fungus for protection. The fungus gets glucose via photosynthesis taking place within the algae. The algae benefit by being protected from the environment by the fungus, which gathers moisture and nutrients from the environment, and anchors the lichen to the ground. A cynic might say that this is not a mutually beneficial collaboration (i.e., not a true symbiotic relationship) but an example of one life form (the fungus) trapping another to increase its own survival. Since we can't interview the algae to find out if they are happy with this arrangement (this takes us back into metaphysics), we can't really know. All we can say is that lichen can survive in environments that neither algae nor fungi alone can, and these very slow-growing lichen can be extremely long-lived (over 100 years).

Importantly, once lichen started breaking up the bare rocks into soil, through both their fungal tendrils and the secretion of acids, other species were able to venture out of the water, like the "pioneer" fungus, Tortotubus. This subterranean fungus (living under the soil surface), which survived for about 70 million years, recycled dead plants and thus made possible the conversion of the land from simple crusty green films to a tropical tree-filled forest (50). Tortotubus did this in part by converting nitrogen-containing compounds in dead plants (and later in dead animals) back into nitrogen and oxygen in the soil.

To better understand this conversion of nitrogen I have to take a step back and explain how plants get the nitrogen (N_2) they need to make amino acids (essential to generate proteins) as well as DNA and RNA bases (to make nucleic acids). Although nitrogen is the most abundant gas in our atmosphere, making up almost 80% (by number of molecules or volume), plants can't use nitrogen gas directly to make amino acids and bases. N_2 must first be "fixed" by bacteria. To do this, the roots of plants secrete chemicals into the rhizosphere (the area around the plant root) to attract specific bacteria. These bacteria help plants grow by converting (fixing) nitrogen (N_2) into ammonia (NH_3), nitrites (NO_2) and nitrates

(NO_3), which plants can then take up and use to make amino acids and nucleic acid bases.

300 million years ago – Oxygen levels peak in the atmosphere

Remember I mentioned that cyanobacteria changed the world by secreting oxygen. Interestingly, this increase in atmospheric oxygen (O_2), from both cyanobacteria and the growing abundance of plant life, likely reached a peak about 300 million years ago (in the Permian age), as determined by the level of rust at this time. So why did this peak occur? It is likely because of a molecule called lignin (meaning "wood" in Latin). Lignan has a ring-like structure (i.e., is a polyphenol). It first appeared about 350 million years ago in an ancestor of modern trees and made it possible for trees, which had a maximum height of about 2 feet at the time, to grow to the height we see today. This new ability to grow tall is because lignan provides both mechanical stiffness and waterproofing to the cell wall (the latter facilitating the transport of water from the roots to the leaves). Since trees are always competing with each other for sunlight, this gave trees with this new molecule a tremendous advantage by allowing them to outgrow their neighbors.

Interestingly, when lignin first appeared in trees there were no fungi or bacteria that could break it down and so when these first tall trees died and fell, microorganisms could not decompose (oxidize) them. Millions of years later, bacteria and fungi evolved enzymes that could break lignan down, and the process to do so required oxygen. Therefore, before these enzymes appeared, atmospheric oxygen was not being depleted as rapidly and this led to a slow buildup of oxygen levels to almost twice (i.e., about 35%) what they are today (21%)! This high level of oxygen allowed insects, which depend on holes (spiracles) in their exoskeletons to bring oxygen directly into their tissues, to grow to twice the size they are now (imagine dragonflies as big as eagles!) (51). This increase in size can be attributed to the fact that the mitochondria within the cells of all multicellular organisms need oxygen to generate ATP. The more oxygen, the more ATP and this made it possible for organisms to grow larger (which, as I have mentioned before, gives them a tremendous selective advantage).

Also of note, some scientists think the buildup of dead, non-decomposable trees during this time, which is called the Carbonifereous (i.e., coal bearing) period, is responsible for the large stores of coal currently buried in the ground. While there is no argument that huge coal deposits did occur during this period, not everyone thinks it is because of the delay in the ability of microorganisms to break down lignin (52).

270 million years ago – Warm Bloodedness

We humans would not exist if our ancestors had not developed warm bloodedness. However, it is important to know that cold-blooded animals, which include all reptiles (i.e., snakes, lizards, turtles, tortoises, alligators, and crocodiles), some insects, amphibians (frogs, toads, and salamanders), as well as most fish, are not literally cold-blooded but rather their internal temperature corresponds to the surrounding (ambient) temperature. So, if it is hot out, these animals can move faster than when it is cold. A simple rule of thumb is that enzyme reaction rates double for every 10°C increase. This means that a cold-blooded animal can theoretically react twice as fast at 30°C than at 20°C.

Interestingly, mammals and dinosaurs/birds may have initially evolved warm-bloodedness as a defense against fungal infections because very few fungi can survive in warm-blooded animals. In contrast, cold-blooded animals are plagued with fungal infections. Some scientists put the appearance of warm-bloodedness just before or after the greatest mass extinction of all time, the Permian-Triassic mass extinction, which occurred about 252 million years ago (53). Most evolutionary scientists today think that mammals and late dinosaurs/early birds developed warm-bloodedness independently. If this is the case, it would be an example of convergent evolution, where a similar trait pops up in two different species independently. Another example of convergent evolution is the complex eyes that vertebrates and cephalopods (squid and octopuses) developed independently. Cephalopod eyes can be considered superior to our human (and all other vertebrate) eyes because their nerves are in back of their retinas rather than in front, so they do not have a blind spot like we do.

So, what was responsible, molecularly, for this shift to warm-bloodedness? This brings us back to our old friend and nemesis, mitochondria. Remember the huge evolutionally advantage that our single-celled ancestors acquired when they engulfed the bacteria that evolved into mitochondria, via the generation of much more ATP? Well, if you home in on the electron transport chain (see **Fig 19B**) and specifically the ATP synthase that generates ATP when it pumps hydrogen ions (protons) back into the mitochondria, that is where the answer lies! The evolution of warm-bloodedness coincided with the appearance of a second type of proton pump in the same mitochondrial inner membrane as ATP synthase (probably by gene duplication of the ATP synthase gene). This new proton pump, that we call an uncoupling protein, pulls protons back into the mitochondria, like ATP synthase does, but it generates heat when it does so instead of ATP. So, the first thing you realize is that when you generate heat, you do not make as much ATP, so you have to eat more. A warm-blooded animal needs the same amount of energy as a cold-blooded animal that is eight times larger and this may have led warm-blooded creatures to be smaller than their cold-blooded brethren (54). To retain as much heat as possible, warm-blooded mammals and late dinosaurs/early birds developed fur and feathers, respectively.

Given everything I have said to this point about warm- and cold-blooded animals it is likely that cold-blooded animals fared better than warm-blooded animals during peaks of global warming and warm-blooded creatures did better during global cooling. This theory nicely segues into natural disasters.

Natural Disasters

There have been at least 5 major mass extinctions: (1) the Ordovician-Silurian, which occurred about 450 million years ago (450 Ma), (2) the Late Devonian (375 Ma), (3) the Permian-Triassic, also called "the Great Dying", which was the worst, wiping out 90 - 95% of all species (250 Ma) (perhaps because of massive volcanic eruptions that depleted atmospheric oxygen), (4) the Triassic-Jurassic (200 Ma) and (5) most recently, the Cretaceous-Paleogene (K-Pg) extinction (66 Ma), when a 10 km wide asteroid hit the Gulf of Mexico, unleashing massive earthquakes, tsunamis, and global dust that wiped out 75% of all species including non-avian dinosaurs. It is possible that the global dust cloud that this asteroid generated cooled the planet enough to slow down the cold-blooded dinosaurs (which had thrived for 160 million years), thus making them easy prey for small warm-blooded mammals. However, a very recent study suggests that non-avian dinosaurs may have actually been warm blooded, challenging this long-held hypothesis (55).

Nonetheless, whatever the reason underlying the demise of the dinosaurs, this asteroid was likely responsible for our very existence. A remnant of this event is a thin layer of sediment called the K-Pg boundary, found throughout the world, which contains high levels of iridium (rare on Earth but abundant in asteroids). Importantly, these 5 major extinctions provided tons of food (provided by the newly dead) and allowed the survivors to take over newly emptied ecological niches. In the absence of predators and competition, survivors were able to bloom and diverge into new species. Then, as time passed and the offspring expanded and formed many new species, competition started up again and whittled down the number of species into those who were the fittest. You can thus think of major extinctions as reset buttons. In addition to these major extinctions, there have been many minor extinctions, ice ages and the coming together and drifting apart of continents that can be considered natural disasters or opportunities because of their influence on the survival of different species. Break time!

Becoming Human

Part 4

In this fourth and final section on Evolution it is time to talk about our 5-million-year journey from tree-dwelling apes to apartment-living humans.

If the entire course of evolution was compressed into a single year, the earliest bacteria would appear at the end of March, but we wouldn't see the first human ancestors until 6 a.m. on December 31st
                                                          Jerry A. Coyne

Since we *Homo sapiens* tend to be most concerned about ourselves, I thought you might be interested in how we came down from the trees to be who we are today. There are many different views on what the specific steps were and the order in which they occurred, so forgive me for making this section highly speculative!

(j) Coming Down from the Trees

Our closest living relatives are chimpanzees, and it is likely we diverged from a common ancestor, perhaps 5 - 6 million years ago in East Africa. What is interesting is that chimpanzees do not have opposable thumbs while our more distant relatives, like gorillas, old world monkeys and lemurs do. This suggests that our hands are more primitive than we once thought and that chimpanzees actually lost opposable thumbs rather than us acquiring them after we diverged (56). This loss may have given chimps (and orangutans) better tree-climbing abilities while retention of opposable thumbs may have been better for ground-living humans (and gorillas). Our opposable thumbs have likely changed somewhat since they first appeared in our ape-like ancestors and are now capable of very delicate manipulations, including construction of many lifesaving (and life destroying) tools.

What prompted our ancestors to come down from the trees is very contentious. Some argue that this descent was because East Africa was transitioning from a forest to a treeless savanna. If this transition did indeed occur, one interesting theory to explain it is that a tectonic shift caused a land bridge to form between North and South America. As a result, warm sea currents could no longer enter the Pacific Ocean from the Atlantic and the resulting cooler temperatures in the Indian Ocean caused the lush green forests of Eastern Africa to slowly transition to savannas. As trees became scarcer, it is possible our ancestors were selected for those who could run safely from one tree to another (51).

(k) Narrowing Hips and Bigger Brains

Whatever the reason was for coming down from the trees, "shortly" thereafter, primates began walking upright. Being bipedal was a tremendous step forward (that was unintentional, but I'll take it) because it freed up our arms and hands to carry things like food or clubs/spears. It also allowed us to see prey and danger better in the tall grass of the savannas. Related to walking upright, the discovery of "Lucy", a pre-human primate who lived about 3.5 million years ago in what is now Ethiopia, was unexpected because she was bipedal (i.e., able to walk on two feet instead of four) and yet had a small skull. This finding was a surprise since, until her discovery, scientists thought that increasing brain size was at least partially responsible for walking upright. Yet, here was evidence that bipedalism preceded bigger brains. Regardless, when we changed from quadrupeds to bipeds, we became less dependent on smell for survival since our noses were no longer close to the ground, and more dependent on seeing and hearing. Consistent with this change, of the more than 800 different genes coding for olfactory receptors (proteins on the surface of neurons within the nose to detect smells) in both humans and chimpanzees, far more of these receptors have been mutated into inactive pseudogenes (genes that do not code for functional proteins) in humans (i.e., over 50%) than in chimpanzees (57).

However, walking upright came with a huge cost. More specifically, running, especially long-distance running, selected for many changes to our anatomy (see Section (d) below). One major change was narrower hips, which resulted in a narrowed birth canal and greater difficulty giving birth. This problem was further compounded by increasing brain size. Our current brains are three times larger than chimpanzees! Furthermore, these brains of ours require 20% of our total energy (ATP), which is more than twice as much as that expended by chimp and other primate brains (so we need to eat more). To adapt to our narrowing hips and bigger brains we selected for smaller, less well-developed (premature) babies that were born totally helpless. This helplessness is in marked contrast to chimpanzee and other non-bipedal primate babies who are born with the ability to clamber onto the hairy backs of their mothers. However, this helplessness may have also had a positive impact since it may have selected for cooperation amongst mothers and close relatives to ensure an infant's survival (i.e., "it takes a village"). As well, efficient cooperation amongst women to look after babies, and amongst men to hunt in groups, may have selected for the development of speech. In one imagined scenario it is possible our ancestor mothers put down their babies in a safe place to dig up tubers and edible roots. This early separation of mother and baby may have led to the beginnings of speech so that babies, through crying, could alert mothers of danger. This theory has been put forward by Yuval Harari in his beautiful book, Sapiens (58). However, non-human babies cry for their mothers as well, and never develop speech, so this may be a bit of a stretch. Philosophically speaking, once we became bipedal, our eyes were no longer fixated on the ground and we could look up at the sky and wonder what the sun, moon and stars were all about (59). This may have prompted us to start thinking, when we were safe and our stomachs full, about something other than day-to-day survival.

(l) Fire

Close your eyes and imagine for a moment that you are a shivering caveman living about 2 to 3 million years ago. You have just returned to the vicinity of your cave after a lightning-sparked fire ravaged your territory and chased you, terrified, into the nearest body of water for refuge. As you drag your wet body, maybe draped in a smelly animal skin for warmth, back to your cave to look for surviving relatives, you smell the acrid smoke and see smoldering tree trunks and burning branches on the ground. It is still light out and you are starving, as usual. You see a small dead animal with its fur burnt off and thoroughly cooked. You gingerly pick it up. It is still hot, and you sniff it and take a tentative bite. It tastes totally different from animals you have eaten in the past. Pieces are much easier to rip off the body, chew and swallow. You used to chew meat for hours before you could swallow it and now it just takes a few seconds. You carry the burnt animal towards your cave, warily looking out, as always, for predators. It is getting dark, and you suddenly have a thought. You pick up a still burning branch, holding it far from the burning part, to help see where you are going and maybe ward off animals who want to eat you, or at the least, eat your newfound meal. You find your cave and enter the dark, clammy interior. No one is there and you wonder if your mate or children survived the fire. It is getting cold, and you put the burning branch down near the entrance and sit near it to stay warm, even though the trapped smoke is making your eyes tear. After a few minutes you notice the flame is starting to go out and you have another thought.

You take the branch, with its dimming flame, and go back out to find more branches to keep the fire going in your cave. Your heart is pounding as you run down the hill. You can see many little menacing eyes in the still smoldering forest, reflecting back the light from your fire (because of the tapetum lucidum in the back of their retinas, which reflects the light back to the retinas to increase night vision). Fortunately, they seem afraid of the still burning branch in your hand and you make it back to your cave safely. You pile the branches together and see one branch and then another catch the flame from your still smoldering first branch. The warmth is comforting, and you drift off into a fitful sleep, wondering if you will ever see your mate or children again.

While this may be a highly fanciful scenario and may actually have occurred in steps rather than all at once, it was a tremendous and likely critical technological advancement. By "harnessing" fire, it gave our ancestors protection, light, heat and, most importantly, the possibility of cooking plants and animals. Cooking made the food vastly more digestible (so more calories could be obtained per gram of food) and, by killing microbes in and on the food, it slowed rotting and thus gave the food a longer "shelf-life". This "domestication" of fire cannot be overstated. It is likely what made it possible for early humans to survive, thrive, and become the dominant genus on the planet. This occurred despite losing many babies and mothers, during childbirth, because of our big heads and narrow hips.

(m) The Ascendency of Homo Sapiens – The Last Men Standing

The ability to walk upright on two legs, albeit in a lumbering fashion, is thought to have first occurred at least 4.5 million years ago, with the appearance of the now extinct, ape-like Australopithecus (Southern Ape). These ancestors of ours, with their short legs, wide hips, long forearms and permanently "shrugged" shoulders still retained the ability to travel through the trees. Some scientists think it was the selection for endurance running and not simply walking (being bipedal), that transformed these predecessors into the genus, Homo (meaning man). Homo anatomy is characterized by the presence of spring-like legs and foot tendons, shoulders that rotate independently of the head and neck for better balance during running, big buttocks for stability, long legs for huge strides and, of course, narrow hips for faster running (60).

The genus, Homo, is thought to have first appeared about 2 million years ago, originating in East Africa. They then spread in several distinct waves to places with different climates, such as Northern Africa, Europe and Asia (58). Because the climates were different in these different places, these humans slowly evolved into distinct species (via speciation). Fossil records suggest that those who settled in Northern Africa became *Homo ergaster*, meaning working man, because of the tools found with their fossil remains. These humans may have survived for close to 500,000 years.

Compared to their Australopithecus ancestors, *Homo ergaster* had larger bodies, longer legs, narrower hips and smaller jaws and teeth (suggesting a change in diet and/or using fire to cook). Those who migrated to Eurasia became *Homo erectus*, meaning upright man. *Homo erectus* has also been called Java man because many of their fossil remains have been discovered on this Indonesian island. They had a huge brow ridge and a low forehead. These humans may have shed their body hair, typically present in great apes, and developed sweat glands to keep cool under the hot Asian sun. The loss of body hair would have made it more difficult for offspring to hold onto their mothers and thus contribute to the need for cooperation amongst mothers and close relatives to protect their children. It should be noted that some scientists think *Homo ergaster* and *Homo erectus* were one species and that the former may have given rise to the latter. Those who migrated to the colder climates of Europe and Western Asia became *Homo neanderthalensis*, meaning man from the Neander valley (where remains were first discovered). Compared to us, these Neanderthals were far more muscular, had bigger brains and were well adapted to the cold climate of Ice Age western Eurasia. The remains of other human species have also been found, including *Homo rudolfensis* (man from Lake Rudolf, East Africa), *Homo soloensis* (Java), *Homo floresiensis* (dwarves from the food scarce Indonesian island of Flores) and *Homo denisova* (Siberia). There are likely more species, still waiting to be discovered (58).

You and I are called *Homo sapiens*, which means, somewhat immodestly, "wise men". We are thought to comprise the most recent wave that took place just over 150,000 years ago (this timeline may have to be adjusted if older fossils are found). It is thought that within the last 2 million years, several human species co-existed and around 300,000 years ago *Homo erectus*, Neanderthals and predecessors of Homo sapiens were all likely using fire on a daily basis. There is recent evidence that communal cooking fires may have actually started much earlier (61). It is possible that cooking led to shorter intestines. Long intestines are a big energy drain so this shortening may have freed up the energy needed for the development of the larger brains that characterized Neanderthals and Sapiens. When these Sapiens, who looked very much like we look today, started to spread to other regions, there was likely both interbreeding and fighting with earlier wave humans. In terms of interbreeding, 1 - 4% of modern Middle Eastern and European human DNA is Neanderthal (based on comparisons with Neanderthal DNA collected from fossils). As well, up to 6% of Denisovan DNA is present in aboriginal Australians (58). These genetic findings mean, at least at the time of interbreeding, Neanderthals, Denisovans and Sapiens were not quite separate species.

So why are we *Homo sapiens* the only human species left today? It may be because we were more proficient hunters and gatherers and so outcompeted our relatives. Or perhaps we had more sophisticated language, and thus could organize into bigger groups to defeat them in battle. Other theories include climate change and/or that we alone slowly domesticated wolves about 35,000 years ago, selecting for more gentle and obedient ones that evolved into dogs (62). If we indeed were the only ones who domesticated wolves, it would have helped us in hunting and fighting and, perhaps most importantly, in warning us of intruders. If we had an early warning system of an attack by our adversaries, it would have given us a huge advantage. Whatever the reason, there is good evidence that Neanderthals disappeared about 30,000 years ago and, for at least the last 10,000 years, we have been the only human species around.

(n) The Pros and Cons of Developing Agriculture

Before agriculture (farming), most people lived in small nomadic bands of at most a few 100 people, and survived by hunting, fishing and gathering roots, berries and other fruit. About 10,000 to 15,000 years ago *Homo sapiens* began to settle in lush areas close to a water source, like the hill country of southeast Turkey. Interestingly, this switch to farming appears to have happened independently and at different times all over the world and may have coincided with the end of our most recent ice age about 18,000 years ago and the beginning of a globally stable, warm period (58). Once an agricultural site was selected, there was a lot to do, from building "permanent" homes, to clearing the land, to sowing and maintaining selected, wild seeds. While this farming made possible a more reliable source of food, and thus faster population growth, it was a far more backbreaking lifestyle than in the hunter/gatherer days (perhaps evoking the tale of a lost Eden). Evidence for this new, more difficult lifestyle are fossils with slipped discs, arthritis and hernias (58). As a result of the increased workload, women were "encouraged" to have more children to help with the chores. However, because of cramped conditions and poor sanitation, more children were dying at a young age than in their nomadic days. Over time, and in different parts of the world, these farmers selected for more nourishing (i.e., starchier) wheat, rice, corn, potatoes, millet and barley. However, a diet based largely on grains is typically poor in protein, fat, vitamins and minerals and, at that time, was also hard to digest until starchier grains were developed. As well, unlike the earlier forager lifestyle, you could not run away if overwhelmed by an enemy because everything you invested in was now fixed in place and you had to defend it. As a result, there was a lot of fighting and death. In addition, unlike nomadic hunters and gatherers, who had few belongings, other than maybe a spear, a knife, and a meaningful necklace, farmers, with a permanent place to live could not acquire and hoard possessions! With this newfound ownership came trading and, eventually, capitalism, which of course rewards greed and selfishness and gets *Homo sapiens* further from their original culture of working as an egalitarian collective (admittedly, our transition away from nature and toward ownership and greed is very speculative and very much influenced by my own thoughts on this subject).

Lastly, I would like to touch on how leaving a symbiotic relationship with nature for the "safety and security" of a human-made village might have affected our religious beliefs. There is some evidence that early hunter-gatherers believed in animism (the animation of nature), the idea that all plants, animals and even rocks, thunder and lightning are inhabited by spiritual beings that can influence human lives (63). A common thread amongst these early beliefs was that all living things, including humans, were part of one big family. Over time, however, as *Homo sapiens* interacted less and less as equals with nature and sought to control, domesticate or harness it, the thought took hold that we were distinct/unrelated and superior to other living things. Specifically, as farmers settled into their new lifestyle, their gods slowly shifted from being within every living and non-living thing to specific gods who could ensure a good harvest, fertility for themselves and their domesticated animals and protection from invaders if they carried out the appropriate rituals.

Over time, maybe to reduce any guilt at our mistreatment and slaughter of our domesticated animals, we started to believe that only humans had souls and were born in the image of God (I warned you that this section was very speculative and biased).

Might I suggest you reread this chapter before reading the recap since I covered a huge amount of material (just a suggestion, don't hurt me!). I know there is a lot to take in, especially if you have never taken a biology course.

To recap this overly long chapter, life started about four billion years ago when the Earth had cooled sufficiently to have liquid water. All living organisms need an external source of energy, and the first living things may have been chemoautotrophs that lived beside hydrothermal vents at the bottom of our oceans. Without chemosynthesis our Earth would likely still be a lifeless rock. While life is incredibly resilient, existing everywhere there is liquid water, both on and within the Earth, individual species are very transient, with >99% of all the species that have ever lived now being extinct. For a species to survive it needs to get enough energy to produce more (or equal numbers of) offspring than the number of individuals dying. We are here today because cyanobacteria secreted oxygen into the water and, subsequently, into the atmosphere. Our primitive single-celled ancestors engulfed bacteria that could use this oxygen to "burn" glucose all the way to carbon dioxide, generating a lot of ATP (energy). This acquisition of what we now call mitochondria into our single-celled ancestors likely provided the energy needed for the appearance of multicellular life. While DNA mutations are responsible for evolution, competition drives it. Our human ancestors survived in large part by "domesticating" fire. Lastly, transitioning from a nomadic, hunter-gatherer lifestyle to an agricultural lifestyle has allowed us to "prosper" (acquire belongings) and generate enough food to propagate more rapidly. However, it has also distanced us from nature and, perhaps, led us to become greedier and more self-centered.

Biogenesis

Chapter 5

In this chapter I am going to discuss the current theories on how life might have started on Earth.

Many investigators feel uneasy stating in public that the origin of life is a mystery, even though behind closed doors they admit they are baffled **Paul Davies**

I personally like the term biogenesis (literally meaning the start of life), coined by Henry Charlton Bastian around 1870 to denote the generation of life from nonliving materials. However, I must acknowledge that many scientists prefer the term abiogenesis, introduced by Thomas Henry Huxley ("Darwin's bulldog", for vigorously defending Darwin's ideas). Although Bastian and Huxley were both keen Darwinists, they had a falling out: Bastian claimed he witnessed spontaneous generation (life arising from non-life) through his microscope, whereas Huxley argued that life could not start under today's conditions and that science was not advanced enough to replicate this event (64). Bastian also had disputes with Louis Pasteur, who demonstrated microorganisms are not generated spontaneously (64). Because of these arguments with reputable scientists, Bastian lost popularity, and so did the term, biogenesis.

OK. Now that we have addressed this somewhat trivial issue of nomenclature and you have a basic understanding of biology from

Chapters 3 and 4, we can discuss how life may have started on Earth. I have wrestled with how to present this next section in a logical sequence since there are a lot of interconnected topics to cover. After weighing several options, I have opted to organize this Chapter into specific questions, so here we go!

(a) When Did Life Begin on Earth?

I already mentioned that our sun and surrounding planets (i.e., our solar system) formed about 4.5 – 4.6 billion years ago and that life started on Earth "soon" after our planet cooled sufficiently to have liquid water. The word "soon" here is purposely vague. We have evidence of life as early as 3.8 billion years ago (65) but it may have first appeared much earlier. Unfortunately, fossils of soft-bodied creatures are very rare because they don't leave nice impressions like bones and hard shells do. Therefore, it is difficult to pinpoint exactly when life started. Most scientists believe in spontaneous generation, in which life "spontaneously" emerged from non-life. Since the probability of spontaneous generation is so low, the more time between the appearance of liquid water and the first signs of life on Earth, the more probable such an event might have happened. Therefore, knowing the timing is important in theorizing how life began! Some scientists think life may have started within 100 million years of liquid water appearing. Whether this many years is considered a short or long time, is debatable.

(b) Where Did Life Begin on Earth?

Charles Darwin proposed in an 1871 letter to his friend, Joseph Hooker, that the spontaneous generation of life might have taken place in "a warm little pond" containing ammonia, various salts and an energy source (e.g., lightning). His warm little pond idea was generally accepted until the discovery of hydrothermal vents (underwater volcanoes) at the bottom of our oceans. These hydrothermal vents are often mentioned as places where life may have started spontaneously because they spew forth high concentrations of elements currently present in living organisms (like hydrogen, nitrogen, phosphorus and carbon) at very high temperatures (which can act as an energy source). Complicating things a little, there are currently two kinds of hydrothermal vents, i.e., black smokers and white smokers. The black smokers (also called underwater geysers) typically develop where tectonic plates are drifting apart (e.g., at the mid-Atlantic ridge mentioned in Chapter 2, **Fig 7**). The seawater that seeps into the spaces between the plates is heated up by the underlying magma and spewed back out at very high temperatures (350°C to 750°C). The spewed-out seawater is rich in dissolved minerals, including iron and sulfur, which combine to form black-colored iron monosulfide (FeS). This compound, together with other minerals, precipitates out of solution when it contacts the near-freezing seawater to form tall, black chimneys at the bottom of the ocean. Because of the nature of the dissolved molecules that are spewed out from black smokers, the water around these vents is very acidic (pH 2-3). The less common white smokers, which typically have smaller chimneys, are cooler (about 120°C) because they are not heated by underground magma. Instead, they are heated by a process called serpentization, which uses the surrounding rock as a catalyst to generate hydrogen that can react with CO_2 to generate heat, methane (CH_4) and short hydrocarbons. They are white because they are rich in the white colored elements, barium, calcium and silicon. Unlike at black smokers, the seawater around white smokers is very alkaline/basic (pH >10). Currently, thermophilic (heat loving) Archaea live at both black and white smokers and convert the heat, methane, and sulfur compounds into organic molecules they can use to generate energy (via chemosynthesis). Of note, some researchers have reported Archaea around hydrothermal vents that can survive up to 121°C (thought to be the upper temperature limit for life) (66). Others suggest 140-150°C might be the upper limit and some even claim that there is evidence of prokaryotic existence in vents with temperatures reaching 350°C! (67).

Importantly, at both black and white smokers, spewed out elements are present at much higher concentrations than in the open ocean and the heat around the smokers provides energy for reactions to occur without the need for lightning storms. With this in mind, William Martin and his group recently showed that simple elements present at hydrothermal vents can form life-supporting organic compounds at temperatures present at these vents. Specifically, to mimic a hydrothermal vent, they kept water overnight at 100°C in the presence of the nickel iron alloy, awaruite, which is present at hydrothermal vents and can donate electrons to accelerate reactions. Under these conditions, hydrogen gas, which is abundant at hydrothermal vents, combined with carbon dioxide (CO_2) to form the organic compounds formate, acetate and pyruvate—the last being one of the most central compounds of metabolism (see **Fig 19**) (68*).*

On the other hand, there are some scientists who say hydrothermal vents are still not high enough in mineral concentrations for life to start and are going back to the warm little pond idea. Specifically, they believe life started around surface geysers or other areas that alternate between wet and dry cycles. This allows dissolved molecules to reach very high concentrations when the water evaporates. Amongst these scientists, some think that ponds with clay within the underlying mud might be ideal because for life to begin. That is because clay can form sheets like a deck of playing cards with mineral-laden water in between the sheets. These sheets can also provide solid surfaces (to act as catalysts) for chemical reactions to take place. So, the take away is...the location where life began, if it did begin spontaneously, is still being disputed.

But now we get to the **really big question**...

(c) How Did Life Begin on Earth?

This, of course, is the unresolved biggie! Although there

are several theories floating around on how life may have started on this planet 3.8 to 4 billion years ago, they all have major problems. The biggest problem is that even the simplest living organism is breathtakingly more complex than any non-living thing, and thus it is hard to imagine how even the most basic kind of life could have arisen spontaneously from non-living matter. So, let's start by talking about the 3 essentials needed for a living (viable) cell (at least based on how current cells operate): (1) a stable, self-replicating set of instructions on how to make and maintain a cell (could be RNA or DNA or a simpler pre-RNA), (2) a set of catalysts (protein enzymes or RNA ribozymes) for central metabolism that can both generate energy (ATP) and use this energy to convert substrates into useful products, and (3) some sort of compartment to keep useful molecules concentrated (preferably a double membrane so that it can concentrate protons or other ions between the inner and outer membrane for energy (ATP) generation). Generating all 3 of these components at the same time is asking a lot, even if you have a few 100 million years to accomplish it! Some have argued that these 3 required events could have happened independently of one another, and then became associated, but this theory is extremely hard to imagine.

The RNA World Hypothesis

The RNA world hypothesis states that all life on Earth began from a simple, self-replicating RNA or pre-RNA-like molecule, without the need of any additional molecules. This is currently the most popular hypothesis for how life started because it satisfies two of the three minimal requirements for life; (1) a replicating set of instructions and (2) some enzymatic activity (69). In terms of the former, RNA and DNA are unique in that, via complementary base pairing, they can make exact copies of themselves. However, since RNA is typically single-stranded, its replication is more complicated than DNA, requiring 2 rounds of replication to make an exact copy of the original strand, as shown in **Fig 26**. In terms of enzymatic activity, some RNA molecules are capable of catalyzing (promoting) reactions. These enzymatic RNAs are called ribozymes. It is thought that back when RNA first formed, the enzymatic activity of ribozymes aided in RNA replication itself. If you remember all the way back to Chapter 3, **Fig 9,** enzymes have a three-dimensional pocket/groove called an "active site." This active site is where (a) substrate(s) bind(s) and is converted into (a) product(s). Ribozymes form these active sites via intramolecular (within the same molecule) hydrogen bonding between complementary bases. In living cells today, ribozymes play a very limited role, catalyzing the making and breaking of non-enzymatic RNA molecules (via the making and breaking of the sugar-phosphate backbone (see **Fig 11**)) and, as part of the ribosome, catalyzing the joining together of amino acids (catalyzing peptide bond formation).

Supporters of the RNA hypothesis argue that the limited enzymatic activity of RNA that we see today suggests it is a relic of a very early time in evolution. However, going against the RNA hypothesis, the intramolecular hydrogen bonds used by ribozymes to maintain their shape, and thus, their enzymatic activity, are easily broken (melted) well below 100°C. Therefore, if life began near hot, hydrothermal vents, ribozymes likely did not exist. On the other hand, in support of the RNA hypothesis, all present-day cells use the 4 RNA nucleotides as substrates to generate the 4 deoxynucleotides used to make DNA. Thus, one can imagine that DNA may be a modified, more recent version of RNA. The switch to DNA may have occurred because DNA is more stable and less prone to mutations. RNA nucleotides have an OH group on their ribose sugar, whereas DNA nucleotides do not. Removing the OH group makes the polymer much more stable and resistant to UV damage. In addition, RNA uses the base uracil (U), whereas DNA uses the base thymine (T), which is more stable and resistant to photochemical mutations. In support of RNA becoming DNA over time, current cells have an enzyme called reverse transcriptase, which converts RNA into DNA. This enzyme may have been used early on in evolution to convert the hereditary material of life from RNA to DNA. Despite this support for the RNA hypothesis, it is almost impossible to imagine how a replicating RNA can evolve to become even the most primitive intact cell that follows the Central Dogma (DNA to mRNA to protein).

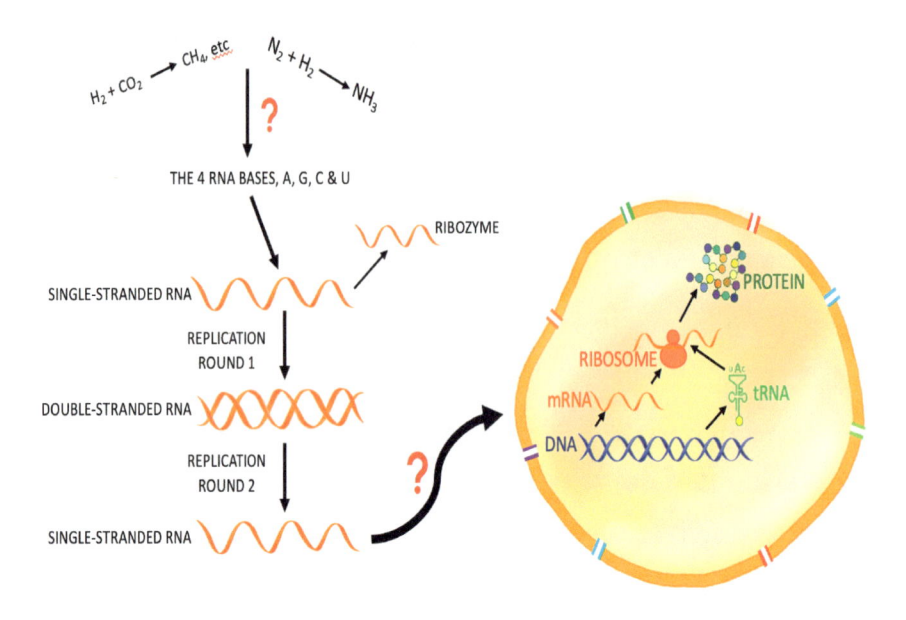

Fig 26. The RNA World Hypothesis. This hypothesis proposes that life on Earth started from a simple, self-replicating RNA molecule. RNA is a single-stranded nucleic acid made up of 4 different nucleotides containing either the base A, G, C, or U. How these nitrogen-containing bases came to be is unknown, hence the first question mark. The RNA hypothesis is popular because RNA provides both a replicating set of instructions and some enzymatic activity via ribozymes. Ribozymes use intramolecular hydrogen bonding to form the 3D structures needed for enzymatic activity. For an RNA molecule to make a copy of itself, it must go through two rounds of replication: round one results in a double stranded RNA molecule and round two results in a single-stranded, exact copy of the original RNA. At some point in evolution, it is imagined that RNA switched to DNA, and then to the Central Dogma (DNA to mRNA to protein). How these primitive cells arouse from RNA is unknown and hard to imagine, hence the question mark.

It is also very hard to imagine how RNA came to be in the first place! The extremely complex, four different nitrogen-containing bases that are needed for RNA (and DNA) synthesis, is unlikely to have arisen spontaneously (see **Fig 26**). On top of that, it is hard to imagine these bases linking together, via alternating phosphate and ribose groups, to form long strands of RNA (see **Fig 13).** Even if these 2 extremely unlikely events did happen, single-stranded RNA is easily hydrolysed (broken down into little pieces), especially at alkaline (basic) pHs, and thus it is hard to imagine generating RNA strands long enough to fold into molecules capable of catalytic activity (70). Because of its pH sensitively, it is unlikely to have formed around white smokers, which, as mentioned above, spew out a very basic pH mix of chemicals. Moreover, RNA's catalytic activity exists because of its ability to form intramolecular hydrogen bonds. However, as mentioned above, these hydrogen bonds are easily broken as the temperature increases and RNA would likely not have had a stable 3D shape and thus no catalytic activity around hydrothermal vents. On the other hand, one can argue that life could still have started at a little distance away from hydrothermal vents, where it is still warm but not hot enough to denature RNA (i.e., break the hydrogen bonds holding them in their unique 3D conformation). However, it is likely the elements spewing out of the vents would not be at a high enough concentration at this distance to make the generation of the RNA possible. So, if you are a proponent of the RNA world hypothesis, you likely have to consider a location other than hydrothermal vents. In fact, because RNA is particularly labile (unstable) at moderate to high temperatures, a number of groups have proposed the RNA world may have evolved on ice, possibly in a eutectic phase (a liquid phase within solid ice) (71). Alternatively, some have proposed a simpler "pre-RNA" as the starting point for life since, as already mentioned, the chance of generating the 4 bases that go into making RNA (or DNA) are too complex to be generated without protein enzymes (72). Another problem is that if you indeed managed to generate a replicating RNA with an RNA sequence that can act as a ribozyme to help it replicate, having a beneficial RNA sequence does not mean, down the road, that this same sequence will get translated into a "useful" protein. In fact, such a possibility is extremely remote. Consistent with this, the ribosomal RNA that currently acts as a ribozyme, attaching amino acids to each other on the surface of ribosomes, never gets translated into a protein.

However, the biggest problem with the RNA hypothesis, at least for me, is that **before** protein synthesis evolved to nudge life towards the Central Dogma, what was the evolutionary advantage of starting to synthesize tRNAs? Remember I mentioned that tRNAs are the "Rosetta stones" of the biological world, translating the RNA language into the protein language). Each tRNA binds a specific amino acid at one end. At the other end, it uses a 3-base sequence (anti-codon) to bind to an mRNA that is being "read" on a ribosome. Thus, a tRNA dictates, via its anti-codon, which amino acid is presented for translation (see **Fig 13**). These "Rosetta stone" tRNAs are not known to have any other function, other than to bring the right amino acids into the right spot on the mRNA for translation into protein. However, they had to appear, evolutionarily, before DNA- or RNA-instructed protein synthesis came into being. While many molecules can be repurposed for other functions during evolution, tRNA does not appear to be one of them. So, why expend energy making tRNAs before protein synthesis reared its head?? Basically, we have a serious chicken (protein) and egg (tRNA) problem with this popular hypothesis of how life started because you can't have one without the other (I feel a song coming on).

The Protein First Model

The protein first model proposes that life originated from self-reproducing proteins. Some believe this model makes more sense than the RNA hypothesis because proteins are far better catalysts than RNA. This is because their 20 different amino acids allow for many more unique 3D shapes than the 4 very similar nucleotides of RNA. As well, at least today, RNA replication itself needs protein enzymes (RNA-dependent RNA polymerases) to generate error-free copies. In support of this protein hypothesis, Stanley Miller and Harold Urey in 1952 exposed a mixture of ammonia, water, hydrogen and methane (to try and simulate the atmosphere that was present 3.5 billion years ago) to a series of electric sparks (to mimic lightning storms). After one week the mixture turned brown and contained over 20 different amino acids (see **Fig 27**) (73). Of note, no nucleotides were generated. This finding was quite exciting at the time, but while their studies convincingly demonstrated they could generate amino acids (the building blocks of proteins), there has been no real progress since these early experiments (with the exception, perhaps, of William Martin's studies mentioned earlier). Moreover, if you remember the Central Dogma, i.e., DNA to RNA to protein, even if these amino acids could randomly self-assemble into some primitive proteins with enzymatic activity (highly unlikely without a protein enzyme or ribozyme that could link them together via peptide bonds), there is no way to go backwards from proteins to RNA or DNA. So, any "successful" protein enzymes (those catalyzing the conversion of substrates into useful products) would not be "remembered" since proteins cannot make copies of themselves, nor can they tell any upstream (RNA or DNA) instruction manual that they are beneficial for survival (and thus be retained for future generations by natural selection of the organism).

To counter the argument that proteins cannot self-replicate, some protein-first proponents argue that some primitive "replication" of proteins might have taken place on solid catalyst-like surfaces under water (74). However, once again, even if this extremely unlikely event occurred, it would not be retained without RNA or DNA to code for this protein sequence. As well, protein synthesis, at least currently, is not only the most energy-draining biological process around but requires a tremendously complex, sophisticated interplay amongst mRNA, tRNA and ribosomes. Thus, it is very difficult to imagine how an early, simpler intermediate stage could come into being spontaneously. While recent studies suggest that some RNA molecules can bind tightly to amino acids (72), that does not explain, as I mentioned earlier, the evolutionary motivation/benefit in generating tRNA, or mRNA and rRNA, to make a protein that may or may not give any advantage to an early life form. An even more recent study suggests that some unusual nucleosides (called non-canonical nucleosides), that are currently present in tRNA and rRNA, and may be relics/vestiges of life's beginnings, are capable of becoming "decorated" with small peptides (several amino acids, linked together in a random fashion). This finding opens up the possibility of actual RNA-peptide complexes that then co-evolved into what we have today (75). While this concept is interesting, any "beneficial" peptides generated in this fashion would occur randomly and thus not be encoded/remembered within the RNA sequence. Also, even the shortest biologically active proteins today are over 50 amino acids long (insulin, for example, is 51 amino acids long), and it is difficult to imagine generating such long peptides via this decorating mechanism.

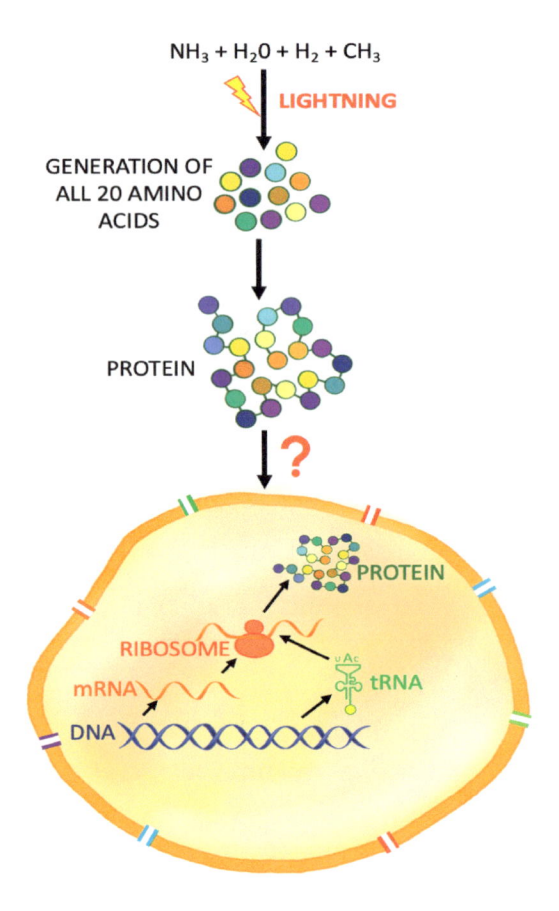

Fig 27. The Protein-First Model. This model is based, in part, on experiments showing that molecules likely present in the Earth's early atmosphere (NH_3, H_2O, H_2, CH_3) were able to generate all the 20 amino acids present in current proteins (but no nucleotides) when stimulated with electric sparks. This model suggests that life started when these amino acids bonded together to form early proteins that could catalyze reactions. However, since proteins cannot replicate or go backwards (cannot code for RNA or DNA), any useful proteins that arose spontaneously would not be retained for the next generation. How early proteins could have led to the first cells, made up of DNA, RNA, and a cellular membrane, is unknown, hence the red question mark.

Membrane First Models

Some scientists argue that some type of compartment/enclosure was the first thing needed for life to begin, so that specific molecules could be concentrated sufficiently for reactions to occur. The membrane for this early compartment is thought to involve an "oily" molecule that is amphipathic; likes water at one end (where it has a polar or charged group) and hates water at the other end (because it is composed of a non-polar hydrocarbon chain) (see **Fig 28**). Proposed amphipathic molecules include phospholipids (existing under water at mineral surfaces) (76), fatty acids (self-assembled and stabilized into bubble-like structures with the help of primitive amino acids. See **Fig 18** for its chemical structure) (77), or simple oil/water interfaces (78). This is a great idea except phospholipids, oils and fatty acids are normally derived only from living organisms. This caveat is not trivial in a lifeless world with no fat-like molecules in it. Where would these amphipathic molecules come from in an abiotic (in the absence of life) world? Related to this, although not generally accepted, Thomas Gold in his book *The Deep Hot Biosphere*, proposed that methane and longer chain hydrocarbons may have been generated abiotically in the Earth's mantle and claimed this is the main source of all hydrocarbon deposits in the Earth's crust. A more likely possibility for the generation of methane and longer chain hydrocarbons abiotically "is at hydrothermal vents pouring out carbon dioxide (CO_2) and hydrogen (H_2) at high temperatures

(300°C to 700°C) in the presence of metal catalysts like chromium (79). This reaction of CO_2 + H_2 to produce CH_4 (methane) at hydrothermal vents is intriguing because it also generates heat (it is exothermic) and so could generate energy for other reactions to take place. However, for these long chain hydrocarbons to form a bubble-like membrane barrier to compartmentalize substrates, they first have to tightly (covalently) link up with a polar or charged group so they can be suspended in water. An interesting alternative that might alleviate some of the problems associated with the membrane first model is that fatty acids have been shown to be present in meteorites! So, it is possible that, back when the oxygen-free Earth was being bombarded heavily by meteorites and asteroids, fatty acids became available for the construction of the first compartments (80).

A problem with these early membranes composed of only phospholipids, oils, or fatty acids is that, while they would allow oxygen and carbon dioxide to move in and out via simple diffusion (which requires no energy), they would not allow any water-soluble molecules like RNA, DNA or glucose to enter (which today require dedicated protein channels, see **Fig 28**). So, how would these first membranes ever go from simple empty bubbles to housing water-soluble molecules like RNA or DNA? To get around this problem, some membrane-first enthusiasts have argued that the appearance of simple membrane "bubbles", replicating molecules (i.e., RNA) and metabolic catalysts (e.g., proteins) arose independently, around the same time, and then came together. If this is correct, then these bubbles would have had to be able to open and close at different times (i.e., be in a dynamic equilibrium between these two states). Although this idea of a dynamic membrane is possible, it would also result in the loss of "cell" contents during opening phases.

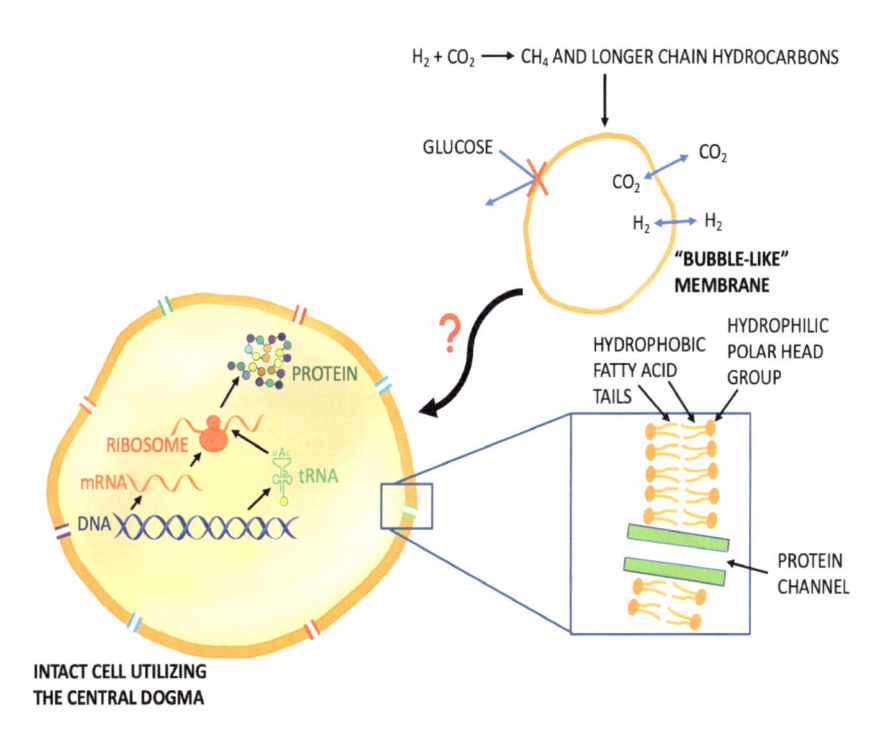

Fig 28. Membrane First Model. The membrane first model suggests that the first step towards life on Earth was a simple "bubble-like" membrane composed of fatty acids, phospholipids or other amphipathic molecules. Amphipathic molecules are hydrophilic (water-loving) on one end, and hydrophobic (water-hating) on the other. The hydrophobic portions are made up of hydrogen and carbon molecules, like methane (CH_4) or longer chain hydrocarbons. It is possible that CH_4 and hydrocarbon chains originated on Earth abiotically at hydrothermal vents, from carbon dioxide (CO_2) reacting with hydrogen (H_2). The hydrophilic portions of amphipathic molecules make up the exterior and interior

of the "bubble-like" membrane, where they can interact with the watery milieu. The hydrophobic portions of amphipathic molecules, hide away from water, making up the inside of the membrane. Because of this hydrophobic interior, large hydrophilic molecules, such as glucose, cannot pass through the membrane, whereas small molecules like CO_2 and H_2 can pass through via simple diffusion. To enable large hydrophilic molecules entry and exit from the cell, present day cell membranes have protein channels that let specific water-soluble molecules in or out (see enlarged insert). How these early "bubble-like" membranes evolved into present day cells, containing DNA, RNA, and protein, is unknown, hence the red question mark.

Metabolism-First Models

Another theory is that the first "biological" molecules on Earth were formed by underwater, metal-based catalysts. While catalysts can act like enzymes to help convert certain substrates into "useful" products, metal-based catalysts are typically not very specific and so this would lead to a very haphazard series of reactions. Since an energy source and a rich concentration of elements would also be required, under-sea hydrothermal vents are likely sites for metabolism-first reactions to occur. In some ways, metabolism-first models are attractive in that they, in theory, can generate the nucleotides (building blocks) needed for RNA or DNA synthesis. However, metabolism first models are difficult to imagine without a compartment to ensure high enough concentrations of substrates and products. Also, it is hard to imagine metal-based catalysis leading to something as specific and complex as pre-RNA, RNA or DNA.

So, looking at the different theories mentioned above, one might conclude that "you simply can't get here from there." However, there is one more, albeit currently fringe, model to consider:

Panspermia (Pansporia)

Panspermia is the idea that life, or the building blocks of life, were seeded into the universe from somewhere other than Earth and made their way to our planet, perhaps via comets, asteroids and/or meteorites. Although "spermia" in the word, panspermia, means "seed," it sounds similar to sperm. Therefore, the word panspermia might lead some to imagine sperm fertilizing some "egg" when it reaches a goldilocks planet. To avoid this confusion, I prefer and will use the term pansporia. Pansporia (panspermia) was first proposed way back in the 5th century, BC, and has some illustrious proponents, including the Swedish Nobel laureate Svants Arrhenius, Sir Fred Hoyle, C Wickramasinghe, Leslie Orgel and Nobel laureate Francis Crick (of Watson & Crick fame) (81). While it has garnered little traction amongst most evolutionary biologists, I think it is worth serious consideration and I will outline the reasons below.

1. Proponents of pansporia are divided into two camps, the

more conservative being that life's building blocks (nucleotides, fatty acids and amino acids) were carried to Earth by meteors, etc., and the more extreme being that intact microbial life itself was carried to Earth. In support of the first camp, amino acids, fatty acids and all five nucleotide bases (adenine, guanine, cytosine, uracil, and thymine) that make up DNA and RNA have recently been detected in meteorite samples (82). One must always bear in mind, however, that meteorites can get contaminated after reaching Earth from terrestrial soil, which contains all of these molecules.

2. An argument for the more extreme camp is that there is evidence that certain single-celled life here on Earth can survive in outer space for long periods of time. Aggregates of Deinococcus bacteria, for example, with their thick cell walls, have recently been shown to survive outside of the International Space Station for at least three years (the experiment was stopped after 3 years)! Deinococcus has been nicknamed "Conan the Bacterium," because of its impressive resistance to acid, UV radiation, extreme temperatures and dehydration, and has been detected floating 7.5 miles above Earth's surface! It has been proposed that certain single-celled prokaryotes can last an almost infinite amount of time if shielded within rock crevices, for example, within asteroids. This type of pansporia within rocks has been called lithopanspermia. Related to lithopanspermia, millions of kilograms of micrometeorites reach the Earth's surface every year (83). A team of prominent scientists at MIT and Harvard have designed an instrument that can be sent to Mars to look for DNA or RNA, in hopes of finding evidence in support of pansporia. This instrument is called The Search for Extraterrestrial Genomes, or SETG.

3.   Some bacteria today (not many, just some members of the Gram positive Firmicutes phylum) are able to form endospores. Endospore formation (or endosporulation) is usually triggered by a lack of nutrients or exposure to radiation. In response to these stressors, the bacterium duplicates its DNA and asymmetrically divides into two daughter cells. However, both daughter cells stay within the original cell wall, so it is a pseudo-cell division (see **Fig 29**). One daughter cell then engulfs the other! The engulfer is referred to as the mother cell, and the engulfed is called the forespore. This engulfment creates an additional cell membrane around the forespore (see **Fig 29**). Next, the mother cell's DNA gets degraded and she works together with the forespore to enclose the latter in a cell wall, and then, on top of that, a thick cortex, a spore coat and, lastly, an exosporium. The forespore then becomes completely dehydrated and released as a mature endospore. Once released, it can remain dormant, purportedly for millions of years, and is amazingly resistant to extreme temperatures and harsh chemical treatments. Once favorable environmental conditions return, the endospore swells, ruptures its thick protective coat and comes back to "life" (84). Endospores preserved in amber for 40 million years have been revived! The reason I am mentioning

endospores is that I think an endospore-like organism might be responsible for pansporia.

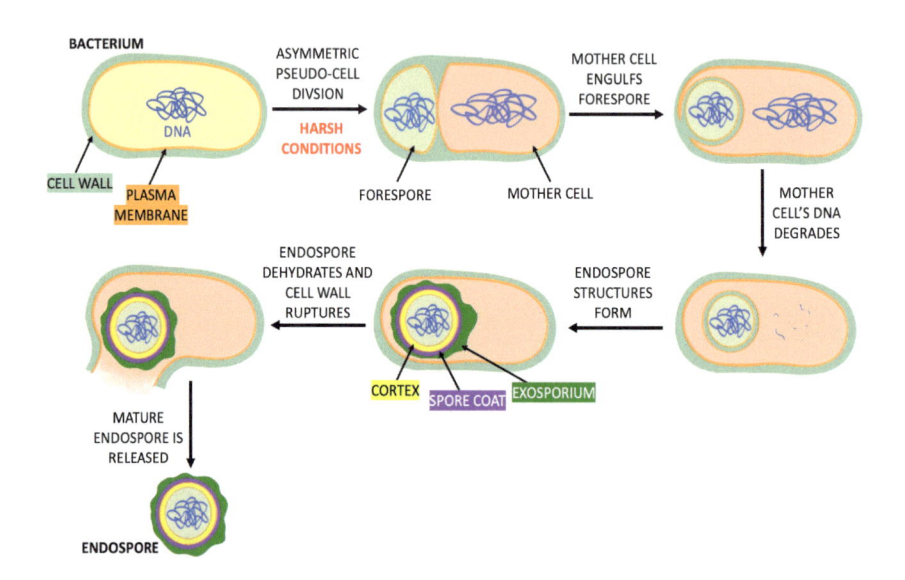

Fig 29. Bacterial Endosporulation. In response to environmental stressors, such as nutrient deprivation, specific types of bacteria can develop into endospores. First, the bacterium makes a copy of itself and asymmetrically pseudo-divides into two daughter cells. These daughter cells are kept in close proximity, surrounded by the original bacterial cell wall. The larger daughter cell is called the mother cell and the other smaller one, the forespore. The mother cell engulfs the forespore, surrounding it in two cellular membranes. Soon after this engulfment, the mother cell's DNA degrades. At this point, the forespore is called the "core" and is surrounded in the mother's cell membrane (the core membrane). Around the core membrane a cell wall develops, followed by a thick cortex, a spore coat, and an exosporium. All these structures together make up the endospore. Next, everything becomes completely dehydrated, and the cell wall of the original bacterium ruptures to release the endospore. Endospores are the most starvation resistant biological structures known. They have no enzymatic activity because they are completely dehydrated (enzymes only work in water). Once

favorable environmental conditions arise, the endospores swell and rupture, to release a revived and functional bacterium.

4.   Mycoplasmas are one of the smallest bacterial-like, free-living single-celled organisms. They have the smallest known genome and are parasites that infect plants and animals. Mycoplasmas lost their cell walls and many biosynthetic systems over time because they were either not needed or they were provided by their host. Nevertheless, these minimal organisms have all the essential genes needed for DNA replication, transcription and translation. Importantly, scientists, including Craig Venter's Genomics group, showed (by inactivating one gene at a time) that in one species of mycoplasma, *Mycoplasma genitalium,* 375 of their 470 genes are absolutely essential for survival (85 – 87). In other words, if you lose any one of these 375 genes, the cell ends up missing an essential protein and dies.

The reason I am bringing up these mycoplasma studies as supporting evidence for pansporia is they reveal that even the simplest life **today** requires hundreds of genes to function/survive. This finding, of course, does not mean the first living organism had to have 375 genes. However, if the first living organisms transitioned to using the Central Dogma sometime in their evolution, they likely required over 300 specific genes, coding for 300 specific proteins, to do this. Thus, it is extremely difficult to imagine spontaneously generated life, via either self-replicating RNA, protein, metabolism or some compartment, alone, or in combination, getting to this required number of genes and proteins.

So, what would it take to prove that pansporia was actually responsible for life on Earth? Strong evidence would be if we find "life" on another planet and, importantly, that this life adheres to the Central Dogma (DNA to RNA to protein). For example, if we found "life" in underground pockets of water on Mars or on one of the 92 known moons of Jupiter. Related to this, while it is thought that Mars and Earth formed about the same time (4.5 billion years ago), Mars is about ½ the size of Earth and, being smaller, cooled sooner and lost its liquid metal core and thus its magnetic field. As a result, the solar winds likely blew away its atmosphere. Despite these harsh conditions, however, there is underground water on Mars (88) and, as mentioned earlier, anywhere on Earth where there is water, there is life. Thus, if the idea of pansporia pans out (nice alliteration there!) it is possible that life will be found underground on Mars. However, even if we find such life, and it employs the Central Dogma, being scientists, we will have to say this is *consistent* with spores populating the universe, rather than absolutely proving it since it is conceivable life always randomly ends up following the Central Dogma (highly unlikely).

At this point you might say, "OK, if you are proposing that life arose on Earth via pansporia, aren't you just kicking the can down the road?" And yes, you would be right. If life on Earth came from elsewhere, this still doesn't explain how it arose in the first place. Life arising spontaneously on another planet raises the same problems as it does on Earth, regardless of the unique conditions of that planet. So, I know this will be very contentious, but I am proposing that carbon-based life in the universe did not start spontaneously, on any planet, but was rather synthesized by sentient beings. I am further proposing that these beings were not carbon-based since this would just get us caught up in another chicken-and-egg situation. While it is impossible for us to imagine the motivation of these potentially long-dead sentients to cobble together "seeds" to populate the universe, it is fun to speculate. Perhaps they were a race of beings that were dying out and wanted to keep life going or were/are immortals that did not want to be alone in an empty universe, or they were a life form that was non-mutating and thus unable to evolve. Perhaps this potentially long-dead species sent out spores simply as an experiment to see what would happen to them on different planets, each planet having unique attributes and challenges (this of course appeals the most to me, being a scientist).

While it is impossible to answer any of these questions, I would urge caution if pansporia turns out to be true and, at some future date, we encounter an extraterrestrial life form that, like us, evolved from a universal spore. I say this based on what we know about ourselves, and that all current life forms on Earth arose, for the most part, via ruthless competition. Also, please do not misconstrue what I am writing here as some belief on my part in a supernatural being or god. I am merely proposing that a different life form from us may be responsible for bringing life, as we know it, to Earth.

So, given everything we have talked about to this point, let's do a little thought experiment. What would you say if I asked **you** to design a single-celled organism that could "seed" life into the universe? I know this is a somewhat leading question, given that I have programmed you to some extent with everything I have covered. Nevertheless, after some reflection, you might say that we need to make something that could remain viable for extremely long periods of time under extremely harsh conditions (i.e., in the presence of cosmic rays and the extreme temperatures of outer space). As well, you would want something that would not start growing until it hit a Goldilocks planet. So, you may want to make a dehydrated, capsule-protected type of cell that would not rupture/grow until it hit water. Then, you might want it to be able to mutate/change to optimize itself to the environment(s) of the planet it lands on, and then somehow remember changes that were beneficial so the offspring would have a better chance of surviving. It would also be beneficial if it could use the most abundant elements in the universe to get started. So, making it a chemoautotroph capable of combining hydrogen (the most abundant element in the universe) with CO_2 (spewing from volcanoes and hydrothermal vents) to make both organic material and a little energy might give it a good start. If I was interested in evolution, to see what advanced organisms might arise on a planet, I would make survival dependent on the ability to get energy from their environment. This dependency would lead to competition (and, at times, cooperation) and thus select for those organisms most capable of acquiring energy. It is certainly interesting that all these qualities exist to varying extents in current life forms on Earth. One, of course, could counter that all the attributes listed above that would support interstellar survival would also be advantageous to life forms that spontaneously developed on Earth and had to survive transient severe droughts and loss of nutrients. And you would be right. I am not saying pansporia **is** how life started on Earth. While it is a hypothesis I currently entertain, based on everything I have mentioned, if new experiments demonstrate that, for example, replicating RNA or pre-RNA can lead to a cell capable of becoming a functional chemoautotroph, I will change my mind (such is science).

To recap, life likely started on Earth about 3.8 – 4.0 billion years ago when the planet cooled enough for liquid water to appear. Although there is no consensus on **where** life started, theories range from hydrothermal vents deep under our oceans, to salty little ponds on the Earth's surface, to water trapped within ice. There is also no consensus on **how** life started but the most popular theory at the moment involves replicating RNA or pre-RNA with some primitive catalytic activity. I propose, on the other hand, that pansporia be given serious consideration, specifically the extreme form involving the dissemination of endospores capable of "coming" to life upon encountering water.

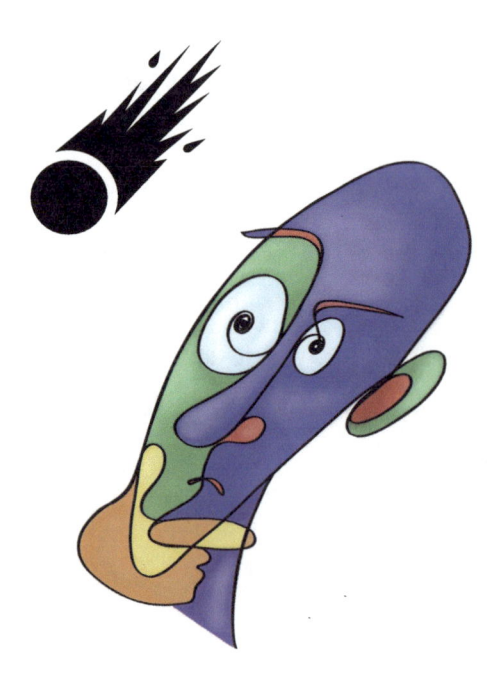

Final Thoughts

Chapter 6

In this final chapter I would like to share my thoughts on where I think we humans are heading and where I think we should be heading.

More than any other time in history, mankind faces a crossroads. One path leads to despair and utter hopelessness. The other, to total extinction. Let us pray we have the wisdom to choose correctly **Woody Allen**

Woody Allen once said, "I don't want to achieve immortality through my work, I want to achieve it through not dying." Unfortunately, since living forever is not yet possible, I thought I would write this book to gain a little immortality before senility robs me of the opportunity. So, first of all, thank you for being curious enough to read this whole book! It's certainly weird to think, as I type this on my computer, that you may be reading this long after I'm dead, but such is the transience of life. So why am I writing this? Well, of course, vanity is a factor, and it is always nice to get some pats on the back before I go. However, I am mainly writing this book because I think we humans have amazing potential to be kind, wise and loving and I'm frustrated with how we keep letting ourselves down. I also know that our decisions, both big and small, should be based on knowledge and compassion rather than surges of testosterone and behaviors that served us well during our evolutionary scramble to survive, but not anymore. So, after more than 50 years as a biochemist studying life, I decided to write down some of my thoughts.

I am finishing this book, after 13 years of part-time writing, in the depressing year of 2023 AD. This is an emotionally debilitating time, with COVID19 still raging 3 years after it started, taking with it an avoidable, world-wide death toll approaching 6 million people. On top of this pandemic, there are the increasing, man-made, climate change-induced weather extremes which threaten our very existence. Compounding these two events is Vladamir Putin's invasion of neighboring Ukraine. This ridiculous war has resulted in the diversion of taxpayer money from education and the prevention and curing of diseases into arms manufacture, and to increased hatred amongst us and totally needless killing on both sides. As if that were not enough, there is the ongoing astonishing level of greed that is increasing the disparity between rich and poor around the world as well as the explosion of misinformation-fueled hate and fear. The war, in part, has also led to fossil fuel shortages and skyrocketing inflation, increasing an already intolerable burden on the growing underclass. All of these things have led to a hard-to-shake malaise that makes it very difficult to be optimistic about the future of mankind. Making things worse is the fact that the global population is continuing to increase and has just surpassed 8 billion people. This fact, coupled with the ongoing, climate-change induced increases in fires, floods and poor crop yields, makes it very likely that we will soon find ourselves in tribal skirmishes for diminishing natural resources.

On the other hand, to provide a glimmer of hope, there have been many such periods of malaise in the past. One of the worst in recent history occurred at the end of World War I when men, both physically and emotionally crippled, returned home in 1918, totally disillusioned with human behavior (often including their own). Making matters worse, they brought home the "Spanish" flu, which then gutted the human race for two years, killing 20 to 50 million people around the world! So, this current "end-of-the-world-as-we-know-it" feeling is not unique, and we still have a chance to reverse things. Importantly, I do not think we can solve the mess we are currently in by praying to a hypothetical supernatural caretaker. **We have to take responsibility for what we have done to this planet and decide if we care enough to save it.** Will short term greed win out? Perhaps. The future of this wonderful planet is in our hands and our hands alone! A few brave and innovative people may yet be able to bring us back from the brink. Sometimes I think of us as "pre-humans", with amazing potential. As Carl Sagan, a wonderful man in so many ways, once said: *For small creatures such as we the vastness is bearable only through love.*

And I agree completely. The only way out of the current mess is for each of us to sincerely care about **all** humanity, not just ourselves, our kin or our tribe. All human beings are basically the same. We all just want the best for our children unless we get twisted by environmental factors, such as abusive relatives, devastating wars and/or punishing poverty. It is also important to remember that all living things today are survivors of 4 billion years of evolution. That is certainly something to celebrate but, as I mentioned already, some of the genes/behaviors that got us here are now wreaking havoc. It is time to grow up, which means wanting everyone on this little blue planet to feel loved and appreciated. And I would extend this love to all living things for, as I mentioned earlier, we are all one big, amazingly complex family, each of us programmed to survive as best we can. One last note on this subject is that we may wipe out humanity if we don't change course, but life will still go on. Some species will adapt no matter how the Earth changes, based on all the natural disasters that have occurred during evolution. I don't know if that is of any comfort to you.

The second to last thing I want to say is that we humans are capable of understanding our surroundings through observation, experimentation, and our capacity to reason. This ability, in itself, is quite remarkable and has taken us out of the caves and into the relative safety of our current lives. The ability to reason has also led to remarkable scientific breakthroughs in understanding how the human body works and how to repair it when it gets damaged. So, I feel we should be celebrating science and what it has done, and can do in the future, to make our lives better. If you or a family member are thinking of embarking on a career in science, do not be intimidated, thinking you have to be a genius to become a scientist.

The most important characteristic you need to be a good scientist is curiosity. It is this curiosity that pushes us to understand the world around us. Science doesn't claim to know the truth, but it certainly tries to figure it out. Scientific "truths" are continuously changing as more information becomes available, through the dedicated researchers who have toiled and continue to toil to understand this universe we live in.

    Lastly, I would sincerely like to thank you for having the patience to read all this and I hope some of it has been informative. Go gently into the future and try to make this world a kinder and better place for everyone!

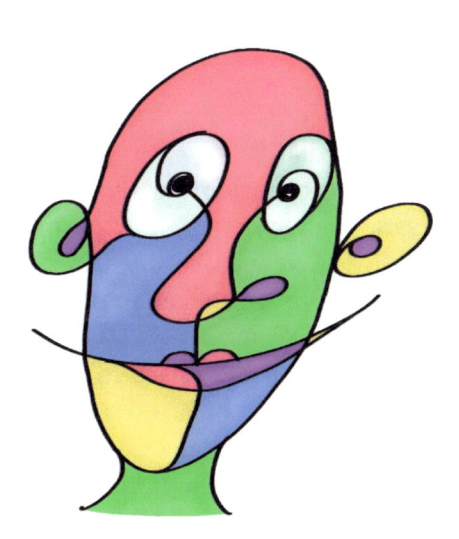

References

Chapter 1

1. Wood, C. "Big Bounce Simulations Challenge the Big Bang". Quanta Magazine. Aug 5, 2020.
2. Glendenning, NK. "Our Place In The Universe". March 21, 2007.
3. Sagan, C & Tyson, ND. "Cosmos: A spacetime odyssey" 2014.
4. Pasley, J. "Nuclear fusion: how scientists can turn latest breakthrough into a new clean power source". The Conversation Dec 14, 2022.
5. Mack, K. "The End of Everything". 2020
6. Bjerkeli, KC. "Naked Science. Birth of the Earth" U-Tube.
7. High School Chemistry/The Dual Nature of Light. Wikibooks. Dec 14, 2021.
8. Patil, V. "How do you measure the distance to a star?" Science ABC. March 15, 2023.
9. Betts, J. "Mass vs weight: simple breakdown of the differences". March 2, 2021.
10. "Graviton". Wikipedia.
11. "Gravity". Wikipedia.
12. Bryson, B. A short history of nearly everything". 2003.
13. Lea, R. "Dark matter could be a cosmic relic from extra dimensions". LiveScience. April 9, 2022.
14. Crew, B. "No, the universe is not expanding at an accelerated rate, say physicists". Science Alert. Oct 24, 2016.
15. Andrei, C, Ijjas, A and Steinhardt, PJ. "Rapidly descending dark energy and the end of cosmic expansion. PNAS, 119, #15. Feb 19, 2022.
16. "Cosmological theories through history". The physics of the universe.

Chapter 2

17. "Greenhouse effect". Wikipedia.
18. "Topsoil". Wikipedia.
19. "Plate tectonics". Wikipedia.
20. "Tectonic motion: Making the Himalayas". Nature. PBS. S29 EP8: The Himalayas Feb 11, 2011.

Chapter 3

21. Uddin, F, Rudin, CM and Sen T. "CRISPR gene therapy: Applications, limitations, and implications for the future". Front Oncol. 10:1387, 2020.
22. Gribaldo, S & Brochier-Armanet, C. "The origin and evolution of Archaea: a state of the art". Philos Trans R Soc Lond B Biol Sci. 361:1007, 2006.
23. Geddes, L. "Fear of a smell can be passed down several generations". New Scientist. Dec 1, 2013.
24. Yang, J-H. et al. "Loss of epigenetic information as a cause of mammalian aging". Cell 186:305, 2023.
25. Tacchi JL. et al. "Post-translational processing targets functionally diverse proteins in *Mycoplasma Hyopneumoniae*". Open Biol. 6, 2016.

Chapter 4

26. Webb, SH. "How Darwin's wife saved his theory". First Things. May 29, 2009.
27. Suzuki, Y et al. "Discovery of life in solid rock deep beneath sea may inspire new search for life on Mars". News Release. April 2, 2020.
28. Gershman, SJ et al. "Reconsidering the evidence for learning in single cells". eLife Jan 4, 2021.
29. "Cleve Baxter". Wikipedia.
30. Khait, I et al. "Sounds emitted by plants under stress are airborne and informative". Cell 186:1328, 2023.

31. Appel, HM & Cocroft, RB. "Plants respond to leaf vibrations caused by insect herbivore chewing". Oecologia 175:1257, 2014.
32. Liester, MB. "Personality changes following heart transplantation: The role of cellular memory". Med Hypotheses 135:109468, 2020.
33. Abramson, CI & Chicas-Mosier, AM. "Learning in Plants: Lessons from Mimosa pudica. Front Psychol. 7:417, 2016.
34. Osimo, EF et al. "Inflammatory markers in depression: A meta-analysis of mean differences and variability in 5,166 patients and 5,083 controls. Brain Behav Immun. 87:901, 2020.
35. Petersen, JM et al. "Hydrogen is an energy source for hydrothermal vent symbiosis". Nature 476:176, Aug 10, 2011.
36. Alberts, B et al. "Electron-Transport chains and their proton pumps". Molecular Biology of the Cell. 4th edition. Garland Science 2002.
37. Souba, WW & Pacitti, AJ. "How amino acids get into cells: mechanisms, models, menus, and mediators. J Parenter Enteral Nutr. 16:569, 1992.
38. Fazekas, A. "Mystery of Earth's water origin solved". National Geographic. Oct 30, 2014.
39. Gronstal, A. "Life in the extreme: hydrothermal vents. Astrobiology at NASA. Nov 5, 2021.
40. Brouwers, L. "Did life evolve in a 'warm little pond'"? Scientific American. Feb 16, 2012.
41. Dubach, I. "Earliest signs of life: Scientists find microbial remains in ancient rocks". Phys Org. Sept 26, 2019.
42. Aiyer, K. "The great oxidation event: how cyanobacteria changed life". American Society for Microbiology. Feb 18, 2022.
43. Kasting, JF & Siefert, JL. "Life and the evolution of Earth's atmosphere". Science 296:1066, 2002.
44. Heidt, A. "The long and winding road to eukaryotic cells" The Scientist. Oct 2022.
45. "How do protists reproduce?" Biology Wise.

46. Johnson, SK & Johnson, PT. "Toxoplasmosis: recent advances in Understanding the link between infection and host behavior". Annu Rev Anim Biosci. 9:249, 2021.
47. Brunet, T & King, N. "The origin of animal multicellularity and cell differentiation". Dev Cell. 43:124, 2017.
48. Ratcliff, WC et al. "Experimental evolution of multicellularity". PNAS. 109:1595, 2012.
49. "Cell fate determination". Wikipedia.
50. Smith, MR. "Cord-forming palaeozoic fungi in terrestrial assemblages". Botanical J of the Linnean Society. 180:452, 2016.
51. Sagan, C & Tyson, ND. "Cosmos: A spacetime odyssey". S2E9. The lost worlds of planet Earth.
52. Nelsen, MP et al. "Delayed fungal evolution did not cause the Paleozoic peak in coal production". PNAS 113:2442, 2016.
53. Benton, MJ. "The origin of endothermy in synapsids and archosaurs and arms races in the Triassic". Gondwana Research 100:261, 2021.
54. Rezende, EL et al. "Shrinking dinosaurs and the evolution of endothermy in birds". ScienceAdvances 6:#1, 2020.
55. Wiemann, J et al. "Fossil biomolecules reveal an avian metabolism in the ancestral dinosaur". Nature 606:522, 2022.
56. Almecija, S, Smaers JB and Jungers WL. "The evolution of human and ape hand proportions". Nature Communications 6:7717, 2015.
57. Gila, Y, Man O and Glusman G. "A comparison of the human and chimpanzee olfactory receptor gene repertoires". Genome Res 15:224, 2005.
58. Harari, YN. "Sapiens. A brief history of humankind". McClelland & Stewart, 2014.
59. Sagan, C & Tyson, ND. "Cosmos: A spacetime odyssey". S1E1. Standing up in the Milky Way.
60. Bramble, DM & Lieberman, DE. "Endurance running and the evolution of *Homo*". Nature 432:345, 2004.
61. Irving, K. "Fossilized fish teeth could be earliest evidence of cooking". The Scientist. Nov 14, 2022.
62. Shipman, P. "How humans and their dogs drove Neanderthals to extinction". Harvard University Press. May 15, 2017.

63. Peoples, HC, Duda, P and Marlowe, FW. "Hunter-gatherers and the origins of religion". Human Nature 27:261, 2016.

Chapter 5

64. Strick, J. "Darwinism and the origin of life: the role of H.C. Bastian in the British spontaneous generation debates, 1868-1873". J Hist Biol 32:51, 1999.
65. Mortillaro, N. "Oldest traces of life on Earth found in Quebec, dating back roughly 3.8 billion years". CBC News. March 1, 2017.
66. Martin, W et al. "Hydrothermal vents and the origin of life". Nature Reviews Microbiology 6:805, 2008.
67. Porterfield, A. "How thermophilic bacteria survive, Part II: DNA". Bite Size Bio. July 9, 2016.
68. Martin, WF. "Older than genes: the acetyl CoA pathway and origins". Front Microbiol. 11, 2020.
69. Saito, H. "The RNA world hypothesis". Nature Reviews Molecular Cell Biology 23:582, 2022.
70. Bernhardt, HS. "The RNA world hypothesis: the worst theory of the early evolution of life (except for all the others)". Biol Direct 7:23, 2012.
71. Vlassov, AV et al. "The RNA world on ice: a new scenario for the emergence of RNA information. J Mol Evol. 61:264, 2005.
72. Alberts, B et al. The RNA world and the origins of life". Molecular Biology of the Cell, 4th edition. Garland Science. 2002.
73. "Miller-Urey experiment". Wikipedia.
74. Cepelewicz, J. "Life's first molecule was protein, not RNA, new model suggests". Quanta Magazine. Nov 2, 2017.
75. Muller, F et al. "A prebiotically plausible scenario of an RNA-peptide world". Nature 605:279, 2022.
76. Spustova, K et al. "Subcompartmentalization and pseudo-division of model protocells". Small. 17: Jan 14, 2021.
77. Cornell, CE et al. "Prebiotic amino acids bind to and stabilize prebiotic fatty acid membranes". PNAS. 116:17239, 2019.

78. Trevors, JT. "Possible origin of a membrane in the subsurface of the Earth". Cell Biology International. 27:451, 2003.
79. "Abiotic formation of hydrocarbons by oceanic hydrothermal circulation". Earth-logs. May 1, 2004.
80. Lai, JC-Y et al. "Meteoritic abundances of fatty acids and potential reaction pathways in planetesimals". Icarus. 319:685, 2019.
81. "Panspermia: Encyclopedia of physical science and technology" (3rd edition). Science Direct. 2003.
82. Oba, Y et al. "Identifying the wide diversity of extraterrestrial purine and pyrimidine nucleobases in carbonaceous meteorites". Nature Communications. 13:2008, 2022.
83. Strickland, A. "Bacteria from Earth can survive in space and could endure the trip to Mars, according to new study". Space Science. Aug 26, 2020.
84. "Endospore". Wikipedia.
85. Glass, JI et al. "Essential genes of a minimal bacterium. PNAS" 103:425, 2006.
86. Razin, S. "The minimal cellular genome of mycoplasma". Indian J Biochem Biophys. 34:124, 1997.
87. Liu, W et al. "Comparative genomics of Mycoplasma: analysis of conserved essential genes and diversity of the pangenome". PLoSOne 2012.
88. Houser, K. "New evidence discovered of underground water on Mars". Freethink. July 4, 2021.

ABOUT THE AUTHOR

Dr. Gerry Krystal is a Distinguished Scientist at the Terry Fox Laboratory at the B.C. Cancer Research Centre and a Professor in the Department of Pathology & Laboratory Medicine at the University of British Columbia. He obtained his PhD in biochemistry from McGill University. His research interests are currently focused on the role that diets play in cancer prevention and treatment. He has published over 200 peer-reviewed scientific papers.

ABOUT THE ILLUSTRATOR/EDITOR

Samantha F. Krystal obtained her B.Sc. in Microbiology and Immunology in 2022 from the University of British Columbia. She is a professional artist, who specializes in portraiture and contemporary dance. In 2019, she graduated from the 4-year post-secondary dance training program, Modus Operandi. She currently teaches dance and is part of a Vancouver based dance collective called Okams Racer.

Manufactured by Amazon.ca
Bolton, ON